Al Gore and mainstream
scientists have spent 25
years maintaining that
the good old days were
colder, snowier, and
better. *Dumping snow
into the river after a
blizzard, New York, 1899.*

The measurement of sea ice at the poles using satellites has only been taking place since 1979, during which time Antarctic ice has increased. Before satellites, the ice was measured by sea captains looking nervously past their ships' bows. *The Terra Nova, McMurdo Sound, Antarctica, circa 1911.*

DON'T SELL YOUR COAT

Surprising Truths About Climate Change

HAROLD AMBLER

Lansing International Books

Published in 2011 by Lansing International Books
Copyright © 2011 by Harold Ambler

For information, address
Lansing International Books
P.O. Box 1143
East Greenwich, RI 02818-9998

Library of Congress
Cataloging-in-Publication Data is available.

ISBN 978-0615569048

Acknowledgements

This book could not have happened without the answers to my questions and other assistance by the following (in alphabetical order): Habibullo Abdussamatov, Syun Akasofu, Von Allen, Brady Ambler, Caitha Ambler, Cheryl Ambler, Christian Ambler, David Ambler, Dennis Ambler, Heather Ambler, Read Ambler, Jerry Ashenbrenner, Joe Bastardi, Robert Blaha, Radu Boghici, Pål Brekke, Tony Brown, Nigel Calder, Dave Carnoy, Bob Carter, John Christy, John Coleman, Benjamin Cook, Piers Corbyn, Bar Yohai Davidoff, Gorm Dybkjær, Don Easterbrook, David Easterling, Jack Eddy, Christopher Essex, Florence Fetterer, Chip Fletcher, Cecily Fong, Valerie Fong, Ze'ev Gedalof, Carlo Giraudi, Albert Hafner, Alan Hamlet, Dierk Hebbeln, Mark Jacobson, Evan Jones, Phil Jones, Thomas Karl, Chris Landsea, Dennis Lettenmaier, Richard Lindzen, William Livingston, John Lyman, Ryan Maue, Walt Meier, Andrew Montford, Rick Moody, Nils-Axel Mörner, Steven Mosher, Timo Niroma, Randall Osterhuber, Eugene Parker, Tim Patterson, Matt Penn, David Peterson, Margo Pfeiff, Eric Posmentier, Bruno Prior, Andy Revkin, Nir Shaviv, Cal Smith, Willie Soon, Roy Spencer, Leif Svalgaard, Henrik Svensmark, Ian Plimer, Bob Rose, Paul Reiter, Cliff Ollier, Andreas Schmittner, Mark Serreze, Michael Steele, Daniel Sumner, George Taylor, Mitchell Taylor, Peter Taylor, Bob Tisdale, Norbert Untersteiner, Tony Velasco, Richard Vigilante, Anthony Watts, Spencer Weart, James Weyman, and Josh Willis. Thanks to them all. I would like to especially thank my wife, Kim Edge Ambler, for her beautiful design work, extraordinary editorial assistance, and patient support, and my daughters, Annalee and Rell, for inspiring Dad to do his best work.

Arctic sea ice is something of a knit cap, one that frays badly every summer. The broken ice makes trips by research vessels, such as the one from which this photo was taken, possible. *The Arctic Ocean, north of western Russia, 2006.*

Table of Contents

Lingo has long been used to scare off lay people from the lofty realm of climate science. To right this wrong, a modest number of scientific terms, including **interglacial** and **Pacific Decadal Oscillation,** are presented.

There is fundamental injustice embedded in modern environmentalists' efforts to suppress development in the Third World and to roll back centuries of progress elsewhere.

The publication of climate alarmism at *The New York Times* and other media outlets has a long history.

The Arctic and the Antarctic are in the news constantly, and yet the public perception of trends in both locales is fraught with misunderstanding. Two facts: for the last 30 years, there has been cooling at the South Pole and there has been an increase in Southern Hemisphere sea ice.

In the face of withering criticism and outright disdain, a lone voice in Russia dared to question the supremacy of carbon dioxide in Earth's atmosphere. *Habibullo Abdussamatov, the director of the Russian space program's solar science division, 2007.*

Preface

I sometimes wish that we could have, over the next five or ten years, a lot of horrid things happening – you know, like tornadoes in the Midwest and so forth – that would get people very concerned about climate change.

– Nobel laureate Thomas Schelling

I was once an adherent of the view that human activity was dangerously raising Earth's temperature and changing its climate in destructive ways. Comments by scientists like Thomas Schelling, who felt so sure that we were on a precipice that he wanted to see some additional destruction to help him make his point were, to my mind, irrefutable.

Then, in 2008, I happened to come across an article about a solar physicist who was predicting thirty to fifty years of global cooling. This was after I had eaten up literally thousands of stories fed to me by my fellow journalists declaring that humankind was on on a one-way, and very hot, train to Hell. "That's odd," I thought, as I finished reading the article about the obscure scientist. His name was Habibullo Abdussamatov. He claimed that:

Could an obscure Russian solar physicist change scientific history? Possibly. Could he change human history? Possibly. *Abdussamatov outside the Pulkovo Observatory, St. Petersburg, Russia, 2008.*

1. The late twentieth century had been a time of unusually strong solar activity.

2. The global warming on Earth could just about all be attributed to this "solar maximum."

3. Mars had seen its own recent warming, corroborating the effect of the Sun on Earth.

4. Carbon dioxide's ability to alter climate on Earth was exaggerated.

5. The Sun was about to enter a period of relatively low activity.

6. The Earth would begin to cool by 2012.

To say that I found these ideas a little surprising would be fair. I had been pretty well shielded up until then from the evidence presented by global warming skeptics. From that point forward, though, I began to study their work. It turned out that, contrary to what I'd been told by the media, most of these individuals were not on the payroll of Big Oil, unscrupulous, or especially stupid.

Surprising facts about polar bears: (1) They have survived other periods with temperatures warmer than today; (2) the number of bears declined due to overhunting during the 1960s; (3) although their numbers have rebounded since, several hundred are still hunted annually. *Female polar bear, Barter Island, Alaska, 2007.*

1 A Map for the Climate Battleground

The best projections tell us that we have less than 100 months to alter our behavior before we risk catastrophic climate change.

— Charles, Prince of Wales, March, 2009

When I enter conversations with people about global warming, which I do warily, I ask them for proof that it is taking place and for proof that it is being caused by humankind. In most cases, merely asking these questions is considered to be in bad form. Those who are willing to respond to the questions typically appeal to authority: "I know that it's happening because figures in the media and the scientific establishment told me so." When I point out that this is not proof, the individual typically invokes polar bears and the melting of ice caps, two things that I used to worry about myself.

It's at this point in the conversation that I mention that the sea ice around Antarctica, since it started being measured by satellite in 1979, has *increased* and that the record maximum value for the ice occurred in 2007. If I'm feeling courageous, I add that polar bear numbers have swelled during the last 40 years, after an intense period of over-hunting. Most people who believe that human-caused global warming is one of the great problems of our age generally cut the conversation short at this point. It is clear to them, from the evidence presented, that yours truly is in possession of some highly annoying pieces of information and probably not to be trusted.

Believe me, I understand how confusing it is to learn of little-known facts like these. This is especially the case with the media trumpeting, daily, catastrophic sea-ice melt in the Arctic basin, more so following 2007's impressive melt season. But 2007, with the record melt and record ice accumulation at opposite ends of the Earth, may not be so unusual at all. We have been examining sea ice from space, using satellites, for a remarkably short period of time. Pronouncements about the significance of trends in the floating ice (decreasing in the Arctic and increasing in the Antarctic) during the last 30 years are best taken with a grain of salt.

And this is part of a bigger picture. Other means of witnessing weather and climate in new ways, in addition to satellites, that make atmospheric phenomena hyper-visible are at least part of the reason why people have come to fear nature again, in ways perhaps not seen for centuries. Some of the other technologies underlying the palpable sense of fear include: video cameras, cable TV, and the Internet. Combined in human consciousness in unprecedented ways, these digital tools have given a lot of people, among them many scientists, the sense that both weather and climate have gone haywire. Thus, when authoritative commentators intone that weather has never done this before, while images of people waist-deep in floodwaters flicker across your TV screen, the *only* moral position becomes to align your sympathies against whatever caused this devastation. Unfortunately, things are nowhere near this straightforward. Weather and climate catastrophes have been killing significant numbers of our species since we first came into existence. Any one of hundreds of vicious storms, for instance, from the last several thousand years, if it were to happen again today and be televised could be taken as proof of "climate change." Despite such a modern lack of perspective, from floods to fires to two-hundred-year droughts, it has already happened before, and, in all too many cases, frequently.

These things didn't only happen in the distant, Biblical past, but merely in the dark void of pre-television. As the last few decades demonstrate, nothing televises so well, or captures the imagination as engagingly, as storms. There is a reason that the average local newscast now contains three or more specific weather segments in a 21-minute broadcast. And the constant air-time that weather now receives, on our iPhones, laptops, desktops, TVs, and cinema screens, has produced a distorted sense that weather used to be benign and has suddenly become menacing.

My own interest in weather has always been intense. When the weather segment came on during the TV news when I was a kid, I would leave whatever conversation I was in, chore I was performing, or the dinner table, to go watch the forecast on Channel 7 from its lead meteorologist, Pete Giddings. That the forecast usually meant, as close to the Pacific Coast as I was, "continued foggy mornings and sunny afternoons" did not dissuade me a bit.

On the other hand, there were occasions when we got less subtle weather where I grew up: Pacific storms, thick with rain and wind, that could and did cause blackouts, clear-sky windstorms, the occasional thunder and lightning, and, in 1974, a snowstorm that left five inches of the most magical substance on Earth in my yard. It makes my stomach clinch up thinking of it, as though for a lost love, 37 years later. Around this same time, my mom taught me a secret: Snow like that at our house that day was frequently just up the road that led to the top of the Santa Cruz Mountains, a drive that she took with my sister, myself, and one friend each on several occasions, carrying our sled in the trunk.

In high school, I still had the sled, and a driver's license. Several times a winter, I drove to the top of the Santa Cruz Mountains, eased myself through a barbed-wire fence, and took the first of many runs down a steep, seldom-used cow pasture hidden from view of the road.

For someone as interested in weather as I, what happened after high school was almost too good to be true: acceptance at Dartmouth College – and complete immersion in the four full-fledged seasons of Northern New England. When the first snowflakes flew sometime in early December of my freshman year, I could be seen in the picture window of Wheeler Dormitory's stairway landing between the third and fourth floor, nose pressed against the glass.

I found a new sport in the new environment: crew. I was soon out on the Connecticut River in all the weather that New England could deliver. More than with a lot of other sports, when your coach tells you that you'll be rowing today, you simply endure whatever comes your way: rain, snow, thunderstorms, wind, frost, sun. When the river was frozen, which it was a lot, we ran and cross-country skied on packed snow paths around town. As a result, we enjoyed winter more than most of our hibernating peers, and my love affair with snow deepened.

Some years later, I came to witness while living in New England again how stressing and dangerous ice, snow, and cold are. Partly my observation came from shoveling my own driveway and those of some relatives and neighbors. Partly it came from editing the obituaries at a newspaper in Massachusetts and taking note of a spurt of deaths during each cold snap. Finally, my observation came from running a woodstove as my primary source of heat for two years. Winter, prior to our modern conveniences, was an ordeal, survived through hard work.

This is what makes it so bizarre that anyone has associated the colder temperatures of yesteryear with prosperity. It has so far not been the case, in human history, that cold times have been good times. And once you come to understand how wrong scientists are on this one point, the question presents itself: What else have they got wrong?

Answering this question requires learning some climate vocabulary. For starters, although people typically discuss global warming in terms of the atmosphere, it is more useful to consider the atmosphere and the ocean together. A term has been developed to facilitate this linkage: the **ocean-atmosphere system**. If you want to warm the atmosphere meaningfully, you have to warm the seas, which cover 66 percent of the planet, as well. That's not easy, given the incredible **heat capacity** of water. Throw a bucket of 125°F water on your body, and it will leave a large red mark. Let a fan blow the same amount of 125°F air on your body, though, and you will barely notice. Water holds a lot more heat than air. On the planetary level, the heat capacity of the entire atmosphere is equivalent to just the top ten feet of the world's seawater. When scientists say that the atmosphere will warm a given amount because of carbon dioxide emissions, they mean that the oceans and the air will both warm, but, again, heating such a water-dominated system is an extremely difficult task.

Despite this difficulty, it does happen. The geologic record is rife with significant ocean-atmosphere warmings, and coolings. One such cool-down, on a major scale, began about three million years ago, as the planet slid into the current **Ice Age**, one of dozens of various lengths that have occurred during the planet's 4.5 billion-year existence. Since the start of the Ice

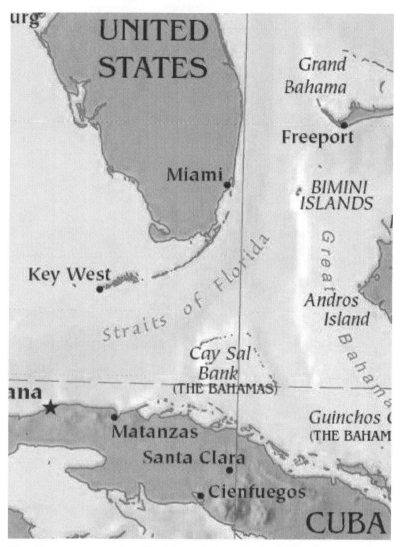

The distance separating Florida and the island of Cuba has varied for millions of years, due to changes in the amount of water captured in ice globally. No single distance is more "natural" or "perfect" than any other.

Age, Earth has been in a deep freeze, with significantly colder oceans than today, for 90 percent of the time. Our modern days are part of a relatively rare respite in the harsh conditions, such interruptions being called **interglacials**.

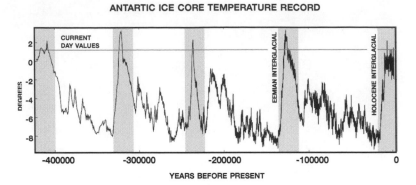

The ice core temperature record from Antarctica shows the relatively cool Holocene interglacial near the right edge of the graph, with the current time represented as Year 0. Temperatures are shown as anomalies, with the Holocene mean temperature taken to be zero. The temperatures of the interglacial prior to ours, the Eemian, about 120,000 years before present, are deemed natural by certain scientists. But any further warming during the Holocene has been presumed by the same scientists to be unnatural.

• •

the cold yields several important consequences, the best known of which are mile-thick ice sheets, covering the greater part of North America, the British Isles, Scandinavia, and a wide strip of Russia. The ice sheets, in turn, have multiple important effects, one of which is removing immense quantities of liquid water from the seas. That is why Ice Age sea level has typically been about 400 feet lower than today. As you would expect, the results of the lower sea level have been extraordinary, from our perspective. Australia has usually been joined by land to New Guinea, likewise England and France. Much of what is now the island realm of Indonesia has been a solid mass that extends to within 600 miles of Australia. The Gulf of Mexico has been far smaller than today; the Florida peninsula has been twice its current width.

Although occupying only ten percent of the Ice Age

The construction of Gothic cathedrals took place during the climactically benign Medieval Warm Period. The cathedral at Rheims, France, exemplifies the opulent display of resources made possible by the era.

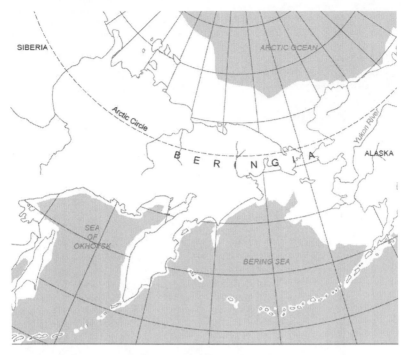

During the next period of widespread glaciation, the sizable region connecting Asia and North America is likely to emerge again. *Beringia, 20,000 years before present.*

timeline, interglacials, normally lasting about 10,000 years each, are scattered in the geologic record at roughly 100,000-year intervals. This means that, in terms of favorable climate, those of us alive in the twenty-first century are about as lucky as we can be. The name of our relatively comfortable climatological nest, which came into existence 12,000 years ago, is the **Holocene interglacial**. It is all well and good for people to argue about what has taken place during the Holocene, which they do, but one thing has to be kept in mind all the while. The Holocene will not last forever. One day, anywhere from ten years from now to 10,000 years from now, the inevitable slide back into

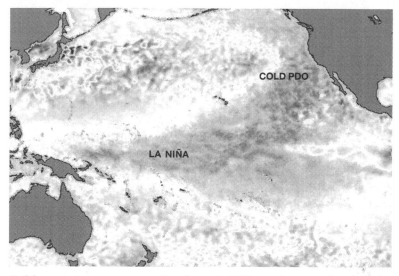

Cold water, in grey, some related to a La Niña and some related to the cool phase of the Pacific Decadal Oscillation, is shown across much of the Pacific Ocean. La Niñas, which last a year or two, are more common during the cool phase of the PDO, which lasts about 30 years. The most recent cool PDO, during which relatively cold water resides on the surface of the Pacific most of the time, began in 2006.

full glaciation will occur. With that return to cold will come new, almost unimaginable, challenges in feeding the world's vastly increased population since the last time the ice sheets were present. One of the changes that the Holocene produced was Chesapeake Bay, which formed when the rising waters associated with glacial melt flooded the Susquehanna River Valley.

The **Holocene Optimum**, a period warmer than our own, started 8,000 years ago and ended 6,000 years ago. Warm periods in Earth's past have been called optimums by climatologists, because the overwhelming majority of life forms on the planet, including human beings, thrive in relatively warm conditions. It happens

25

DEC 97

NASA

The great size of the Pacific is such that relative warming and cooling over the ocean's equatorial waters have worldwide effects. *The Super El Nino of 1997-98, a tongue of warm water shown in white, stretches across the Pacific.*

to have been during the Holocene Optimum that agriculture began to be practiced by our forebears and permanent habitations became increasingly common. After the Holocene Optimum came a cool phase, with less optimal growing conditions, which concluded around 200 B.C. At that time, a warm period known as the **Roman Climatic Optimum** began, and it lasted 500 years. The good growing conditions were a core reason for the Roman Empire's success, and although it would be an exaggeration to say that it all collapsed merely because of a cooling climate, colder temperatures and failed crops were at least partly to blame for the fall of Rome.

Starting in about the year 1000, after this harsh stretch, came a far more forgiving climatological time known as the **Medieval Warm Period.** During the next three and a half centuries, sea ice that had encircled Iceland melted. Vikings sailed for Greenland and began a colony that lasted for 350 years. The arts flourished in Europe, as exemplified by the great Gothic cathedrals of the era. Famines became a rarity, and population grew.

Beginning in about the year 1300 and lasting until about the year 1850, came another cold time – the **Little Ice Age.** Winters were brutal; summers were short; alpine glaciers grew; crop failures became frequent; witch burning was prevalent; wars raged; diseases, typified by the Black Death, claimed whole cities and towns. Ironically, despite its many hardships, the Little Ice Age largely established our understanding of what normal is, in terms of glaciers, sea level, and temperature. For, the glaciers that we assumed to have lasted throughout human history had in fact advanced to their (modern) maximum size towards the middle of the 19th century, only a handful of generations ago.

As for the terms to describe our current ocean-atmosphere conditions, I will introduce a few here. First is a 60-year cycle in the largest ocean on Earth, the **Pacific Decadal Oscillation.** While all of the oceans are significant in determining atmospheric conditions all around the globe, the one that calls the shots the most often on a global level is the Pacific. The most widespread and influential cycle in this vast liquid wilderness is the Pacific Decadal Oscillation. Referred to by experts simply as the **PDO**, the Pacific Decadal Oscillation has a 30-year warm phase and a 30-year cool phase.

The 60-year cycle has a relationship to the better-known phenomena of **El Niños** and **La Niñas**, shorter-term events known by people around the globe to spur both droughts and floods as well as warm and cold temperatures, depending on the part of the world in which you live. For the southwestern United States, El Niños are warm, precipitation-enhancing events, and La Niñas are cool and dry. In the southern and eastern United States, El Niños are cold and wet, and La Niñas are relatively warm and dry.

In the simplest terms, an El Niño is a superficial band of warm water extending along the Equator between Indonesia and Peru. The largest, most intense El Niño measured by scientists formed in 1997 and lasted through early 1998. On temperature-anomaly maps, showing the departure from normal sea surface temperatures, the so-called "Super El Niño" looked like a bright-red tongue extending two-thirds of the way across the Pacific. The band of warm water was so large in size and so warm that it spiked the **global mean temperature**, an average of sea surface and air temperatures around the world, to the highest point that it had ever attained.

El Niños are complex in their genesis. Broadly speaking, they are born in the following sequence of events: (1) water, pushed by westerly **trade winds**, piles up in the eastern Pacific; (2) the vertical Sun over the Equator hits the accumulated water long enough to warm it, creating an enormous, superficial "warm pool"; (3) the trade winds slow down, or even reverse, moving from east to west; (4) pockets of warm water the size of Mexico slosh back toward the western Pacific, in what are called **Kelvin waves.**

La Niñas, conversely, are superficial pools of unusually cool water lying along the Equator. One of the most significant of these took place in 2007 and 2008. That La Niña, which on temperature-anomaly maps resembled a deep blue band stretching two-thirds the way from Peru to Indonesia, briefly cooled the global atmospheric temperature by just about the same amount that it had risen during the previous 100 years.

La Niñas are hatched in more or less the opposite process as the one leading to **El Niños.** Rather than slowing down, the dominant west-to-east trade winds speed up, pushing water towards the west. Cold water wells up along the coast of South America and then moves along the Equator, propelled by brisk trade winds. More cool water is brought to the surface along the Equator by the wind as the top layer of the ocean overturns.

When the Pacific Decadal Oscillation, a cycle arguably underlying American weather journalism for the last century and a quarter, sits in its 30-year warm phase, El Niños are relatively powerful and relatively frequent, and the global mean temperature tends to increase. You can experience a La Niña during a warm PDO, but it is usually a weak one. Conversely, when the Pacific Decadal Oscillation is in its 30-year cool phase, La Niñas are more prevalent. When the 1997-98 Super El Niño occurred, it was about two-thirds the way through the warm Pacific Decadal Oscillation of 1978 to 2005.

It is no coincidence that widespread concern about manmade (so-called anthropogenic) **global warming** got going during the recently ended warm phase of the Pacific Decadal Oscillation. Conversely, talk of an Ice Age was all the rage as the cool phase of the Pacific Decadal Oscillation played out some thirty years ago. This is a point that cannot be overemphasized. As world temperatures have stabilized since 1998, causes of the last century's global warming are finally being debated.

As that debate unfolds, three points are worth bearing in mind: (1) Planet Earth – its ocean-atmosphere system, to be precise – has been cooling since the Holocene Optimum, which ended 6,000 years ago; (2) civilization has flourished during periods of warmth, including and especially during the last 150 years; (3) the time of most extensive glacial coverage during the entire Holocene interglacial was the Little Ice Age, which ended just a century and a half ago.

All of this is suggestive of the simple fact that climate changes, and has always changed. Wishing for it to do otherwise is anti-Nature and anti-science. As the philosopher George Santayana said, "To be interested in the changing seasons is a happier state of mind than to be hopelessly in love with spring."

There is no morally robust argument to keep the undeveloped world in the dark. *Satellite photo of Europe and Africa, 2008.*

2 A Scandalous Timeout

The environmental movement I helped found has lost its objectivity, morality, and humanity. The pain and suffering it is inflicting on families in developing countries must no longer be tolerated.

– Patrick Moore, founder of Greenpeace

How did we get to the point where a chat about climate change makes a conversation about abortion among people from wildly divergent points of view seem like a walk in the park?

For starters, the road to hell is paved with good intentions. Honest scientists with no political agenda whatsoever originated the theory of manmade global warming. Other honest scientists, well-intentioned and with high academic standards, walked in these individuals' footsteps to examine whether the theory was being borne out by facts. In the mid-1970s, however, climate science and public policy began to mix in ways that have compromised the principles of peer review, faculty hiring, academic funding, and the practice of scientific research itself. The menace justifying such excesses, supposedly that of climate change, is, at least to some extent, capitalist democracy itself. If you're a comfortable or even somewhat comfortable citizen in a capitalist state, you would do well to take notice.

I myself live in a climate-controlled 2,000-square foot home on the southern edge of Austin, Texas. When it is 105 degrees Fahrenheit and I want to plunge in some cold water at one of the local streams, lakes, or swimming pools, I do so. I put my daughter in her car seat, get behind the wheel of our minivan, drive for twenty minutes or so, and the two of us jump in the wonderfully cool water. Sometimes my wife or I will fill our daughter's kiddy pool, and the three of us splash around in that.

When I want to watch television, I pick up the remote control, press a button, and, as the screen flickers to life, I feel, for a moment, omnipotent. I used to watch the Weather Channel a shocking amount – less now that it has become heavily politicized – but I still watch TV weather forecasts on one channel or another multiple times a day. Tennis is another thing that I enjoy watching; my sense of the year I'm living in is strongly influenced by the arrival of the Australian Open in January, the French Open in May, Wimbledon in July, and the U.S. Open in August. Seeing the brilliance of the athletes on the screen makes me feel brilliant. I have played fewer than 30 rounds of golf in my life, but this, also, is something that I could watch for long periods of time without getting bored.

When I want to read news headlines, I sit down at one of my computers, push some buttons, and world-class news services appear, as if by magic, on the screen in front of me. If I want to check in with people with whom I went to high school or college, I may do so with a small amount of effort. When I desire a few quick games of chess, I have only to press some buttons, and then, as if by magic, players of my approximate skill level appear in a queue ready to play a game.

What if I have a fond memory of seeing a musician perform, one that I saw back when I had time to go to concerts during my teens and twenties? I can press some buttons, and in ten or twenty seconds, The Clash, The Who, Michael Hedges, or Willie Nelson appear and give a command performance just for me. Sometimes watching one of them perform on the Internet this way is deeply gratifying. Sometimes I wish I were there in person, too.

When I want to exercise, I drive to a rowing club where I belong and take a single scull out on one of the prettiest stretches of water in the whole country, Austin's Lady Bird Lake.

When my wife and I cannot bear being away from our parents, siblings, or grandparents any longer, we buy airline tickets, go to the airport, and are teleported from Texas to California, or to the Northeast. (I am a native Californian, my wife a native Rhode Islander.) When we are too busy or cash-strapped for such a trip, we spend about as long as we'd like talking on the phone to the people back home. We can even do so from anywhere we might be through the miracle of cell phones.

When I have a sorrow or a hope or something that nothing else will cure, I plug in my electric guitar and play a few licks. Either that, or I sit down at my family's 1910 upright Steinway that could use a little work but that makes us oddly happy whenever one of us tickles its keys.

When one of us is ill, we make an appointment to see a doctor, drive there, get seen, get treated, and commence feeling better – often in a single day.

When the weather turns cold, we commence heating the air in our home. They say there's a good chance you live in Texas if you have used your air conditioner and your heater in the same day, and that's about right. Even when a mean-spirited cold front drops down from the northern plains and lets us realize in a hurry that it's not always hot in Austin, the temperature in my home stays within 5 degrees of 70 about 95 percent of the time.

So, suffice it to say, we are blessed.

Do I have to tell you that there are those less fortunate than we?

What of the nearly two billion people on the planet who are living without electricity? These are nearly two billion souls that live without the following: refrigerated food, clean water, modern sewage, refrigerated medicine, air conditioning, non-polluting indoor stoves, central heat, the Internet, cell phones, television, radio, recorded music players, washing machines, dishwashers, and microwave ovens. The suffering created by such want is heavy. It also happens to fall disproportionately on women, who must carry water, and perform other heavy domestic tasks in the darkened locales of Africa, Asia, and South America.

Westerners, in the comfort of their clean, well-lighted dens, like to imagine that if they lived in such conditions, there would be something redemptive about it, perhaps drawing upon their recollection of a week spent at a National Park. But the outdoors, experienced without the filter of modern technology, is: hunger, insect bites, filth, disease, stress, and premature death. That is what the great outdoors are waiting to provide anyone who has any illusions about being *at one with Nature*. With the benefit of modern technology, though, nature suddenly becomes a hauntingly beautiful backdrop to an individual's emotional theater. But this *modern* experience of sunsets, sunrises, scenic animals, wide panoramas, and the rest as something sublime, as a thing to be maximized, stems from the deep knowledge within the individual that he is in control of his circumstances. This control itself stems from the fact that, in virtually all cases, the individual traveled to the land of natural glory using the internal combustion engine, in one form or another.

So, the perceived love of transcendent Nature is, in reality, the love of anti-Nature, of the internal combustion engine itself, with the Superman-like power afforded by the automobile and the modern transportation network.

Someone who makes thorough use of all forms of travel made possible by the internal combustion engine is former Vice-President Al Gore. For his own reasons, he would preserve the status quo, with the developed world remaining developed, and the undeveloped world being stuck in excruciating limbo. In other words, in order to save the world from the perceived ills of climate change, he is comfortable stamping energy inequity on the face of the globe indefinitely. Under the banner of saving Earth from the greenhouse gas carbon dioxide, he and his colleagues have been setting up carbon trading schemes at a rapid clip. These include for-profit initiatives among corporations, as well as intergovernmental efforts that supposedly profit no one. Both, however, lead to the trading of carbon in speculative markets, shadow stock exchanges of sorts, and tens of billions of dollars have changed hands in this way.

Among other negative effects, cap-and-trade would largely erase the hope of development for the Third World. If industrialized nations wish to hit any targets for carbon reduction, they must purchase carbon credits from the governments of nations unable to provide electricity to more than a lucky few of their people. The payment for these carbon credits is unlikely to do much good, as corruption among such governments is endemic, and the likelihood of any trickle-down benefit ever reaching the cold, huddled masses is slim.

Thus, at a time when fully industrialized nations find themselves at a massive advantage in the world economic competition, calls for cap-and-trade should be heard for what they are: a demand for a lengthy – if not permanent – time-out in the development game. Ironically, the self-appointed moral gate-keepers on climate appear heedless of the their brothers and sisters heating and cooking by burning dung in poorly ventilated mud huts and seldom living past the age of 40. This is all to say that the "progressive left" doesn't really have a "progressive" or "left" position on the issue of carbon dioxide. Rather, Al Gore's moral army has taken a position that serves the elite of the best-off countries in the world. Meanwhile, carbon phobia, threatening the very existence of so many people on the planet, is based on science that will be scoffed at within one human generation.

A 1903 Missouri
River flood, besides
impacting travel,
toppled bridges,
killed many, and
left 20,000 homeless.
*Passengers and
crew stand atop the
Missouri-Kansas-
Texas train.*

3 Giving Shape to Phantoms

The most effective way to sell daily news is to make it worrisome.

– Christopher Essex and Ross McKitrick,
Taken by Storm

Get it out of your head: weather didn't used to be friendly. It didn't used to rain just enough, snow just enough, with the wind blowing just enough, and the Sun shining just enough. Things didn't recently go to Hell in a hand basket. That is just a story. And it's not a particularly hard story to prove false.

On the other hand, reciting the story in public, like a sort of catechism, allows large segments of the population to communicate to one another that they are of strong moral character. The story invariably includes the items on the following list: atmospheric warming, polar ice cap melt, rising sea levels, habitat destruction, drought, fire, flooding, hurricanes, tornadoes, the cessation of the Gulf Stream, and the spread of malaria. One of the reasons that I know this list is that I feared these things for a couple of decades myself, beginning in 1985 when I took a meteorology course in college.

I had been an environmentally conscious, politically liberal person since a young age, besides being a kid obsessed with weather. How my liberalism came about is fairly conventional: an upbringing in a hippie-tinged household in the San Francisco Bay Area during the 1960s and 1970s. One of the first things I can specifically remember writing was a letter to the prime minister of Japan, urging him to end his country's whale hunt. I must have felt strongly about it, and I must have expressed myself well, because my teacher called my mom to tell her that I had done a good job.

As for politics, I was a little-boy Democrat throughout elementary school, successfully betting a classmate in sixth grade that Jimmy Carter would prevail against Gerald Ford. When, four years later, Reagan beat Carter, I took it on the chin. My impression of Reagan at the time was that he was a tough-talking, saber-rattling demagogue who would eventually say or do something that would lead to nuclear weapons raining down on my head.

So, by the time I got to college, I was a liberal with a capital "L," one especially disposed toward rescuing nature from the evils of corporate greed. Still, a meteorology course I took during my sophomore year was an eye-opener. Not only were industrialized nations coughing out horrifying quantities of particulate waste, we learned, leading to acid rain and other ills, but invisible pollution in the form of carbon dioxide was likely to raise Earth's temperature calamitously. From that point forward, I read everything I could about

If the Dust Bowl were to happen again today, it would be televised 24/7 and attributed to climate change. *Left: an abandoned farm, near Liberal, Kansas, March, 1936. Right: farm flanked by new sand dunes in Oklahoma, April, 1936.*

manmade global warming, and worried about it a fair bit.

I was unaware, in my angst, that among all the consequences of manmade global warming that I had been taught to fear, not one had actually taken place. Initially through the course and then through the newspaper and magazine articles that I read, my reality had been eroded and then re-created by the output of computers. It is not surprising that I had begun to confuse the output of computers with physical reality, because most of the news articles I was reading at the time regarding climate change were guilty of the very same crime. Sea level could rise in a way that it hadn't before during the Holocene; weather patterns could change catastrophically; deadly floods could become more widespread; but here on the ground, on planet Earth, not one of these things had happened.

Only when I began taking a skeptical look at the hair-raising articles that I had read through the years, in the wake of my newfound interest in solar variability and the work of Abdussamatov, did I come to see that journalists had been hawking climate scare stories for more than a hundred years. If you don't believe me, take a look at a century and a quarter's worth of foreboding climate statements by one of the country's most illustrious publications, *The New York Times*, as an example:

June 23, 1890

The older inhabitants tell us that the Winters are not as cold now as when they were young, and we have all observed a marked diminution of the average cold even in this last decade.

February 24, 1895

The question is again being discussed whether recent and long-continued observations do not point to the **advent of a second glacial period**, when the countries now basking in the fostering warmth of a tropical sun will ultimately give way to the perennial frost and snow of the polar regions.

March 27, 1933

America in **longest warm spell since 1776**; temperature line records a 25 year rise.

May 30, 1947

A **mysterious warming** of the climate is slowly

manifesting itself in the Arctic, engendering a "serious international problem."

January 30, 1961

After a week of discussions on the causes of climate change, an assembly of specialists from several continents seems to have reached **unanimous agreement on only one point: it is getting colder.**

July 18, 1970

The United States and the Soviet Union are mounting **large-scale investigations to determine why the Arctic climate is becoming more frigid,** why parts of the Arctic sea ice have recently become ominously thicker and whether the extent of that ice cover contributes to the onset of ice ages.

December 29, 1974

A number of climatologists, whose job it is to keep an eye on long-term weather changes, have lately been predicting **deterioration of the benign climate to which we have grown accustomed.**

May 21, 1975

Sooner or later a **major cooling** of the climate is widely considered inevitable. Hints that it may have already begun are evident.

January 5, 1978

An international team of specialists has concluded

from eight indexes of climate that there is **no end in sight to the cooling trend of the last 30 years**, at least in the Northern Hemisphere.

January 7, 1982

Mankind's activities in increasing the amount of carbon dioxide and other chemicals in the atmosphere can be expected to have a substantial **warming effect on climate**, with the first clear signs of the trend becoming evident within this decade, a scientist at the National Aeronautics and Space Administration said here today.

February 4, 1989

Dr. Hansen and some other scientists believe **high temperatures of recent years** indicate the greenhouse effect is already occurring.

January 16, 2007

The **water was littered with dozens of icebergs**, some as large as half an acre; every hour or so, several more tons of ice fractured off the shelf with a thunderous crack and an earth-shaking rumble.

April 26, 2009

With diminished rice harvests, seawater seeping into aquifers and islands vanishing into rising oceans, Southeast Asia will be among the regions worst affected by **global warming**, according to a report scheduled for release on Monday by the Asian Development Bank.

No one should get the impression, however, that the *Times* has been alone in speaking about climate from a vantage point of anxiety, if not terror, for quite a while. Here is just a smattering of other doomsday climate predictions about global cooling:

September 10, 1923

The discoveries of changes in the sun's heat and southward advance of glaciers in recent years have given rise to the **conjectures of the possible advent of a new ice age.**   – *Time*

December 23, 1962

Like an outrigger canoe riding before a huge comber, the earth with its inhabitants is caught on the downslope of an immense climatic wave that is **plunging us toward another Ice Age.**

– *The Los Angeles Times*

July 9, 1971

"[Earth] could be as little as 50 or 60 years away from a **disastrous new ice age**, a leading atmospheric scientist predicts."   – *The Washington Post*

June 24, 1974

Climatological Cassandras are becoming increasingly apprehensive, for the weather aberrations they are studying may be the **harbinger of another ice age.**

– *Time*

April 28, 1975

There are ominous signs that the Earth's weather patterns have begun to change dramatically and that these changes may portend a drastic decline in food production – with serious political implications for just about every nation on Earth. The drop in food output could begin quite soon, perhaps only 10 years from now.... The central fact is that after three quarters of a century of extraordinarily mild conditions, the earth's climate seems to be cooling down.   – *Newsweek*

March 6, 1975

A recent flurry of papers has provided further evidence for the belief that the ***Earth is cooling***. There now seems little doubt that changes over the past few years are more than a minor statistical fluctuation.   – *Nature*

July 26, 1999

Scientists are warning that some of the Himalayan **glaciers could vanish within ten years because of global warming**. A build-up of greenhouse gases is blamed for the meltdown, which could lead to drought and flooding in the region affecting millions of people.   – *The Birmingham Post*

November 1, 2000

Officials with the Panama Canal Authority, managers of the locks and reservoirs since the United States relinquished control of the canal in 1999, warn that **global warming, increased shipping traffic and**

**bigger seagoing vessels could cripple the canal's
capacity to operate within a decade.** – CNN

Unfortunately, very little about climate science is as simple as any of these journalistic pronouncements suggest, but he temptation to deliver a scoop, and to sell newspapers by whatever means necessary, has been irresistible. Some similar motivation, surely, led the producers of *An Inconvenient Truth*, and Al Gore himself, to strive to convince an entire generation of intelligent people that the rising quantity of carbon dioxide in the atmosphere was a menace to their very existence. Theatergoers felt it had to be thus, having seen a graph showing temperature increasing with carbon dioxide levels throughout the geologic record. Unfortunately, this is simply backward. It is CO_2 that has responded to temperature changes throughout the geologic record. The two are separated by an 800-year lag, and it is the *temperature* that leads. As news of this fact became more widely known in the wake of questions raised by Gore's film, scientists in the scare-mongering camp changed their tune: Periods of warmth are *prolonged*, they now maintain, by carbon dioxide increases, even if the gas does not *cause* the original warming. Meanwhile, the great majority of the public still erroneously believes the central theme of Gore's film: that carbon dioxide *caused* the temperature increases in the recent geologic record. The facts refute this supposition.

Among the reasons for carbon dioxide's relatively small

effect on atmospheric temperature is that the gas can absorb only a limited amount of infrared radiation. The reason for this is that CO_2 absorbs heat only along limited bandwidths, and the bandwidths in which it absorbs are already largely covered by other greenhouse gases, water vapor and nitrous oxide among them. The idea that CO_2 is a "blanket" capturing nearly all heat trying to leave the Earth is false. It has more in common with a heat "sieve," and is up in the atmosphere with other sieves with similar size holes at similar orientations.

Even as many scientists and climate-focused journalists continue to warn of human emissions' dangerous warming effects, it is worth remembering that just a generation ago, scientists and journalists were making the contrary assertion: that a human-caused Ice Age might be nigh. It is ironic, then, looking back on the *Times*' and its peers' various warnings about dangerous, growing cold and dangerous, growing warmth, that there appears never to have been a time when these observers looked at the global mean temperature and judged it to be "just right." Further, these self-professed nature-lovers within the scientific and media establishments are among those who seem to fear the natural variability of climate the most. They promote the idea that before the existence of modern technology climate was benevolent, and that today, perverted by a trace gas that constitutes 0.039 percent of the atmosphere, climate has become a deadly foe.

The way to get more and more people to accept this strange view of climate requires more than drawing attention to a rise in the global mean temperature by three-quarters of one degree Celsius over the course of a century and a half. It requires recasting the whole of weather and climate as a Frankenstein monster.

A *Denver Post* article from 1990 captures this increasingly alarmist view of climate:

> Huge sand dunes extending east from Colorado's Front Range may be on the verge of breaking through the thin topsoil, transforming America's rolling High Plains into a desert, new research suggests. The giant sand dunes discovered in NASA satellite photos are expected to re-emerge over the next 20 to 50 years, depending on how fast average temperatures rise from the suspected "greenhouse effect," scientists believe.

The very notion that increased carbon dioxide and atmospheric warmth are growing deserts worldwide, is in fact, controversial, and there is plenty of evidence to the contrary. Why would that be? One reason is that plants *like* CO_2. A plethora of studies show that **plants and forests are growing faster** than when carbon dioxide was at a more meagerly level in the mid-19th century. Some studies show them to be growing by as much as 25 percent faster. Nearly twenty years down the road from the original *Denver Post* publication, Colorado's High Plains are as they were before. Neither they, nor the Central Plains, have returned to anything like their Dust Bowl-era worst of the 1930s, the hottest decade on record in the United States.

A second reason that the *Denver Post* piece is flawed: The link between warmer temperatures and the growth of deserts is *inverted*. Among the many proofs of this: The largest desert on Earth, the Sahara, expands and contracts in conjunction with Ice Age cycles. When the ice sheets are at their largest and the ocean-atmosphere system is at its coolest, evaporation diminishes, precipitation lessens, and the Sahara grows. Although the Dust Bowl did take place during several years of heat and drought in the United States, on a global scale and on any kind of longer time scale, more heat in the ocean-atmosphere system equals more evaporation and thus more precipitation.

On a global level, when the ice sheets begin to melt from increasing heat in the ocean-atmosphere system, evaporation increases, precipitation increases, and the Sahara *shrinks*. To recap, ice age = (relatively) big Sahara; warm interglacial = (relatively) small Sahara. The same is true for Earth's other great deserts. Anyone who says that the geologic record shows the growth of deserts, globally, during times of relative warmth needs to look a little harder at the record, to put it charitably.

What of rising sea levels? There is a reason that coastal inundation is on the list of red-hot alarmist subjects for both scientists and journalists. For one thing, perhaps thanks in part to the computer-animation graphics employed by Gore, the rise of the world's oceans is conjured as *something sudden*. One moment, in the film, lower Manhattan looks the way it does today, and the next moment it has been submerged. "Greenland would also raise sea level almost twenty feet, if it went," Gore explains. He then goes so far as to invoke the terrorist attacks of September 11. Against a computer-generated video showing the (fictitious) flooding of Battery Park in New York City, his narrative continues:

> Here is Manhattan. Here is the World Trade Center memorial site. And after the horrible events of 9/11 we said, "Never again." But this is what would happen to Manhattan. They can measure this precisely, just as the scientists could predict precisely how much water would reach the levees in New Orleans. The area where the World Trade Center Memorial is to be located would be under water.

Greenland has shown a certain amount of warming, and melting, around its periphery, but it has also shown significant growth of the glaciers covering its central terrain. Also, the melt has been shown to have been greater during the 1930s than at any other time in the past century. More important, Greenland was quite a lot warmer during a period known as the Medieval Warm Period in and around the year 1000, which is why Vikings put settlements there at the time. Did Greenland's ice sheet melt during the Medieval Warm Period? It did not. What about Manhattan? What is happening to sea levels there? According to the National Oceanic and Atmospheric Administration, sea level there has been rising by a little less than 3

For all the scare stories about rising sea levels, Venice is still open for business. *The island city in a satellite photograph taken in 2001.*

millimeters a year since the mid-1800s and the rate of increase is not accelerating.

But why would Al Gore say that melting ice caps are raising ocean levels (or are on the verge of doing so) dangerously, and why would so many in the mainstream media say the same thing? One answer: videotape of pieces of glaciers falling into the sea.

Just as it had been determined more than a decade ago that the average child had seen several thousand shootings on television by the age of 10, so now, too, has the average child seen footage of glacial calving, generally accompanied by a very serious, if not sinister, voiceover, hundreds of times. The reasonable conclusion drawn from such messages is that if global warming can make a piece of glacier shudder and then drop off into the ocean, then it must be pretty bad. This is effective film-making, as *An Inconvenient Truth* shows, but it is appalling journalism. Why? Because glacial calving has taken place for millions of years, as long as oceans and glaciers have tried to occupy the same space. The image of building-size hunks of ice tumbling into the sea is spectacular, there's no getting around it. It is also a process that has taken place every year throughout human history – and long before that. In and of itself, though, videotape of glacial calving is meaningless. Yet because of such footage, any number of my environmentalist friends believe that they know in their bones that our era is one of perilous warming. It is indeed easy to be manipulated by images.

As for melting of ice on land, it was proceeding at a rapid pace during the dawn of human history 7,000 years ago during the Holocene Optimum. As the Roman Empire collapsed and European civilization became a shell of its former self during the Dark Ages, glaciers advanced. During the Medieval Warm Period, they retreated again. Then, during the Little Ice Age, from 1300 to 1850, glaciers grew, eating up farmland and causing destructive floods.

Our sense of what normal is regarding glaciers largely springs from the experiences of our recent forebears! These are people who happened to come into the world and live their lives in the middle of the 19th century. It is simply happenstance that science became interested in whether the world's glaciers were growing or shrinking in the immediate aftermath of the maximum glaciation during the Holocene. If science had developed four hundred years sooner, instead, scientists might reasonably have concluded that increasingly frigid temperatures and growing glaciers were signs that a new ice age was starting.

As an example, both of my grandfathers were born in the 1890s. Their grandfathers were born when villages in the Alps were being bulldozed by advancing glaciers, when Alaskan glaciers filled bays that today are open water. But *now* just happens to be when *we* live. There is no proof that the melting of glaciers during the past century and a half, which has been shown to have taken place dozens and dozens of times in the past, is the fault of our being here on Earth. We could just as easily, you and I, have been born during the last full-fledged ice age from 120,000 years ago until 12,000 years ago, or even during the Little Ice Age. Both were far less forgiving times in which to live than our own. We could have been born during the comfortable, productive Medieval Warm Period, too, when sea ice and glaciers both melted significantly, and population exploded.

Nonetheless, we happen to live in an era when glaciers are, for the most part, shrinking. Again, this has happened dozens of times before in world's past. The coincidence of the glacial melt with our fascination with carbon dioxide has led, sadly, to the people of the world being held hostage by scientists with questionable acumen and misinformed journalists. And Earth's rising sea level, laid at the feet of glacial melt, is the most potent weapon in the hostage-takers' rhetorical arsenal.

Before getting into the science of sea levels, though,

let us stop, briefly, to consider the nature of water. Stability and water are anything but synonymous. "The river was just at this level yesterday, but today it is here," "sea level has risen during the last thirty years," "glaciers have grown," "glaciers have shrunk," "the fire is burning out of control" and "Australia is undergoing a drought" are all statements that simply express the shifting of water around the globe. It is not in water's nature to be stationary. When reading through both mainstream media and scientific articles the governing idea in a great many of them appears to be that nature used to be friendly and that *water used to be predictable.*

This is not the case. The surface of the spinning orb we live on is 71 percent water. "How inappropriate to call this planet Earth," observed Arthur C. Clarke, "when it is quite clearly Ocean." Between the seas, the great river systems, the glaciers, the lakes, and all the water in the atmosphere, our home *is, in its very essence, change.* The idea that by showing the existence of change with respect to water a scientist or anyone else is showing something new, unnatural, or sinister is *profoundly unscientific,* and reveals a lack of appreciation for the mystery, and the destructive power, of water.

Indeed, one of the great ironies of the global warming alarmism printed in newspapers during the last generation is the suggestion that now is a particularly unfortunate time to have been born, climatologically speaking. This is laughable. Earth is in a prolonged Ice Age. Of the last 200,000 years more than 170,000 of them were times of glacial advance and cold and fewer than 30,000 have passed during *interglacials* – the warm and benign pauses between long periods of deep chill.

Our ancestors in Europe and Asia, meanwhile, were skin-wearing cave dwellers shivering through winters until the brief respite of summer came each year. Life was difficult and short. When dramatic global warming brought an end to the last period of glaciation and cold some 12,000 years ago, it set in motion a range of events up to and including the United States' journey to the Moon in 1969. That trip, and all of the technology that preceded it, was facilitated by the warm, safe nest of the Holocene interglacial. When the lunar landing took place it punctuated the obvious: the best time to be born, in terms of climate, was right now. The benign Holocene made it possible for agriculture to be widely developed and for the incredible multiplicity of labor-reducing and life-giving technologies to be brought to bear. People born during the Holocene have won the climate lottery. Weather and climate did *not* used to be better than they are now, and they're not going to get better when the ice sheets reform.

After the Holocene began, the world's oceans rose 400 feet, and there have been various fluctuations of sea levels since then, with a very modest increase over the past several millennia. But the sea has not accelerated in its uphill run, spurred by increased melting of the great ice sheets. The Greenland ice sheet, for one thing, *did not melt* during the balmy Holocene Optimum; the idea that it would melt now, when temperatures are lower than they were then, is anti-science.

So, how did you and I come to exist? How did our ancestors survive these past sea-level changes of more than 400 feet and manage to have children and successfully raise them? They adapted. I know that sounds hard, and kind of scary, but it is what they did. Humankind has been adapting with respect to ever-changing sea levels as long as we have existed. The notion that technologically undeveloped human beings who lived thousands of years ago were at some kind of advantage when it comes to natural adaptation compared to those of us in the twenty-first century, with the large number of powerful tools at our disposal, is ludicrous. If we need to adapt to rising water, *and we will almost surely not have to during this century*, then that is what we will do. Such adaptation has been managed before, or you and I would not be here.

So, what have sea levels been doing since the latter part of the Little Ice Age 200 years ago? They have been gradually rising, on average about a millimeter a year. Two hundred years, two hundred millimeters – less than a foot. There have been warnings that Vanuatu in the South Pacific is being eaten by rising sea levels, but the tide gauge shows no such rise. The Tuvalu islanders have begun to ask for relocation funds from Western nations, but Tuvalu's case is actually quite instructive. A Japanese pineapple grower extracted so much fresh groundwater from the island that sea water began to seep in. The islanders were understandably upset, but the ruining of their water was not caused by a rising sea. An uncritical western press accepted its script from frustrated islanders, and the result was bad journalism. Two pieces from 2001 exemplify this:

> In ten years time, most of the low-lying atolls surrounding Tuvalu's nine islands in the South Pacific Ocean will be submerged under water as global warming raises sea levels.   – CNN

> The authorities in Tuvalu have publicly conceded defeat to the sea rising around them. Appeals have gone out to the governments of New Zealand and Australia to help in the full-scale evacuation of Tuvalu's population.
> – Andrew Simms, *The Guardian*

It is almost ten years later. Are any of the atolls submerged? Not one.

Another famous case of supposed inundation is that of the Maldives. Ironically, Nils-Axl Mörner, a contributing member of the IPCC who measured the sea level, personally, on repeated trips to the island nation, found that in 1970 sea levels in the Maldives *fell* 20 centimeters, more or less all at once, which he attributed to evaporation. Local fishermen confirmed that he was correct in his measurements, pointing out how the contours of their harbors and channels had changed. *Mörner has recorded no increase in the 39 years since then.*

Sea levels, meanwhile, are one case where people

should feel free to use their own eyes. How many of the world's ports have had to be retooled due to rising water? Has construction on any part of the world's coast ended, because of rising water? Has Venice been submerged? The water comes up to the edge of St. Mark's Square there just as it has for hundreds of years. One legitimate worry is that the land underneath the tourist destination is sinking. In geological parlance, Venice is *subsiding*, and steps have been taken to assure that it does not disappear.

Whatever sea level rise has taken during the course of my own life would be unlikely to cause the demise of a single person. Dozens and dozens of tide gauges bear this out. Measuring at Midway Atoll in the Central Pacific, the National Oceanic and Atmospheric Administration found that sea level rose .7 millimeters a year from 1947 to 2006. At this rate, in a hundred years, the oceans would rise a little less than three inches. Again, such modest, unthreatening rises have taken place dozens of times in the past, and will again.

Fire is another expression of the *ever-changing* distribution of water. A wet, cool forest will seldom burn, but not a single forest stays wet and cool forever. What is required to generate a large forest fire is a period of relatively abundant precipitation followed by a period of relatively limited precipitation. Or, in other words, normal meteorological variability.

Fire photographs so well that commentators feel especially free to associate it with global warming. In the Californian west, where I am from, there are wet and rainy seasons. This has been true throughout the Holocene climate epoch. Yet, it is a safe bet that TV screens will light up in July, August, and September with this year's crop of Western fires, and somebody somewhere, watching and listening, will conclude that the brilliant orange and reds were caused by too much carbon dioxide in the atmosphere and the warming that it produced. But have any of the television correspondents studied American fire history or world fire history?

In the 1920s, according to the U.S. Forest Service, fires burned an average of 26 million acres a year. During the 1930s, they burned an average of 39 million acres a year. During the 1940s, the total was 22 million acres a year. During the 1950s, 10 million acres a year. During the 1960s, the total burned was 5 million acres a year. During the 1970s, 4 million acres a year. During the 1980s and '90s, fires again burned an average of 5 million acres a year. *The record for the 2000s shows an annual average of 7 million acres a year.*

Thus, the overall trend for the past century is overwhelmingly in decline. If the number of acres burned increased next year to the levels of the 1920s and '30s, what do you suppose the newspaper headlines would read?

One of the few countries with good statistics on wildfires, Canada has experienced about 5 million acres of forest fire a year. This is despite the fact that it is more sparsely populated than the United States (humans start a high percentage of wildfires) and that it allows nearly all the fires that take place in its northern forests to burn themselves out. The number

Claims that human emissions of greenhouse gases have increased the size and severity of forest fires are not backed up by facts. In the 1920s, forest fires claimed 26 million acres a year in the United States. During the 2000s, the yearly average was 7 million. *Oregon National Forest, circa 1915.*

of acres burned in Canada has not changed, on a five-year running average, since at least 1970, according to the Canadian National Forestry Service.

Although Australia had its most lethal wave of wildfires in decades early in 2009, the Black Saturday bushfire that burned 1.1 million acres, it has been known for both intense droughts and wildfires since the time of its colonization (and long before by the Aborigines). In 1921, Australia's Superintendent of Immigration H. S. Gullett recognized that the continent's climate was a public relations disaster, and he offered an idea about how to handle the problem:

> Many thousands of Australians go abroad every year on business or pleasure. The Commonwealth Immigration Office appeals to every one of them to embark with the resolve that he will on all possible occasions speak well of Australia. Let none of them speak evil. Such words as "drought" should be thrown overboard as the vessels put out to sea.
>
> – H. S. Gullett, as quoted by
> *The Sydney Morning Herald*, 6 June 1921

References to fire and drought abound in Australian literature, and, like the Native Americans in the United States and Canada, Australian Aborigines used fire to shape the landscape in which they lived, regularly burning huge stretches of the island continent.

Australia's tendency toward fire and drought reached a peak during modern times some 35 years ago. In the

1974-75 fire season, according to the Australian Bureau of Statistics, 289 million acres, or *15 percent of the total land area of the continent*, were claimed by wildfires. If 15 percent of Australia burned in a single season nowadays, this would surely be presented as evidence of catastrophic climate change.

Hard figures on the area burned by wildfires globally do not exist, for a variety of reasons. One is that many nations consider fire, and fire fatalities, to be a source of shame and thus do not want any international attention to be brought to the fires that they experience. Another is that changing governments, and forms of government, in much of the Third World make the keeping of statistics difficult, according to Chuck Bushey, the president of the International Association of Wildland Fire. The former Soviet Union used to allow fires in its northern reaches, principally Siberia, to burn out on their own; the new Russia has continued this policy. There has been talk of a study to analyze historic military satellite images of Siberia to see how much may have burned during the past 40 years, Bushey says, but the study has yet to be completed.

So, although alarmists routinely tout an increased incidence and area coverage of wildfire as proof of global warming, the only countries with extensive records going back more than ten years show no upward trend whatsoever. Australia's worst single fire season was more than 30 years ago. The United States shows a *downward trend* over the past 70 years.

Meanwhile, fire is part of nature. Television viewers

and other news consumers may be shown terrifying images now, but the simple truth is that fires in North America claim fewer acres now than they did a couple of generations ago. Will that stop politicians from trading off the human fear of being burned to death? Will that stop people from building homes in known fire alleys, such as the hillsides of California or the Australian bush – with their known wet and dry seasons and tendency toward drought?

No era is without its perils. One could conclude listening to Al Gore and other alarmists that everything used to be peaches and cream, weather-wise and climate-wise, right up until the time that carbon dioxide started getting emitted in meaningful quantities in the years following the Second World War. They have gone on to say that global warming has engendered unheard-of flooding. Consider the following article from July 1, 2008:

> Floods like those that inundated the U.S. Midwest are supposed to occur once every 500 years but this is the second since 1993, suggesting flawed forecasts that do not take global warming into account, conservation experts said on Tuesday.
>
> – Deborah Zabarenko, Reuters

But, as serious as both the 1993 flood and the 2008 flood were, neither was even in the ballpark of the 1927 Mississippi Flood. And that flood was preceded by devastating floods in 1912 and 1922. All three of the latter have in common the terminal disadvantage of having taken place prior to the introduction of the video camera, of course, and prior to the substantial rise in atmospheric carbon dioxide. A list of years with significant flooding on the Mississippi, according to the National Oceanic and Atmospheric Administration, follows:

1718, 1735, 1770, 1782, 1785, 1791, 1796, 1799, 1809, 1811, 1813, 1815, 1816, 1823, 1824, 1828, 1844, 1849, 1850, 1851, 1858, 1859, 1892, 1893, 1907, 1908, 1912, 1913, 1916, 1920, 1922, 1923, 1927, 1929, 1932, 1936, 1937, 1945, 1950, 1957, 1958, 1965, 1973, 1974, 1975, 1979, 1983, 1984, 1993, 1997, 2001, 2002, 2004, and 2008.

It can be assumed that the reporting for the 18th and 19th centuries was spottier than that for the twentieth and twenty-first centuries. Nonetheless, the 1700s and 1800s clearly saw their own frequent and serious floods. An important point here is that human memory is short. Keep this in mind when TV news crews and other reporters present images of "unprecedented" flooding.

Similar efforts to manipulate informed citizens with regard to flood waters take place on a regular basis in Europe. It's a topic that Bjorn Lomborg explores in *Cool It*:

We seem to have a very selective memory of floods, thinking that our age is special. And in a sense it is, but perhaps not in the way we think. In general, casualties due to flooding have been declining in Europe, with large-scale loss of life seen in episodes preceding the nineteenth century, while the twentieth-century death tolls were significantly lower, and deaths in the 1990s were lower still. Flooding has been pervasive throughout our history. All but two of the fifty-six major floods that affected Florence since 1177 (to pick just one example) occurred before 1844.

The Europe of the Little Ice Age was one in which abundant snow, late spring melting, and ice jams conspired to flood the Rhine and other major rivers regularly, killing thousands. The fact that Earth's temperature has moderated during the two centuries since the end of the Little Ice Age would seem, on the face of it, to be a good thing with regard to European floods. More to the point, as both the case of the Mississippi floods and floods in Florence listed by Lomborg make clear, serious river flooding is not a byproduct of industrialization.

Speaking of variable precipitation, the United States Secretary of Energy, the Nobel-prize winning physicist Steven Chu, has expressed strong concern about snowfall in the California mountains. A February 4, 2009, *Los Angeles Times* article by Jim Tankersley fleshed out the secretary's pessimistic sense of California's ability to find enough water for its farmers and private residents in increasingly meager amounts of snow.

California's farms and vineyards could vanish by the end of the century, and its major cities could be

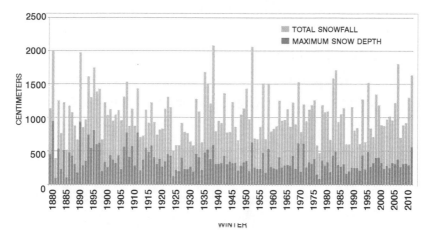

Graph of snowpack, as measured at Donner Summit, California. The most recent season, 2010-2011, saw a top-10 total snowfall and the greatest snow depth in 40 years. Credit: University of California-Berkeley/Central Sierra Snow Laboratory.

in jeopardy, if Americans do not act to slow the advance of global warming, Secretary of Energy Steven Chu said Tuesday....

Chu warned of water shortages plaguing the West and Upper Midwest and particularly dire consequences for California, his home state, the nation's leading agricultural producer.

In a worst case, Chu said, up to 90% of the Sierra snowpack could disappear, all but eliminating a natural storage system for water vital to agriculture.

"I don't think the American public has gripped in its gut what could happen," he said. "We're looking at a scenario where there's no more agriculture in California." And, he added, "I don't actually see how they can keep their cities going" either.

This is egregious alarmism, without any basis in fact.

One has to wonder what kinds of trends Chu is looking at. After a pair of serious drought years in the late 1970s (during which California families recycled dishwater to use in their gardens and took "Navy showers" with the water only turned on for about twenty seconds at the beginning and end of the shower), the Sierra snowpack has returned to a maximum depth of ten feet, with annual snowfall averaging about 30 feet.

The snowpack has declined, minimally, compared to the unusually wet winters of 1935 to 1945, but it has increased compared to the years of 1924 to 1934. Looking at the maximum snowpack depth and annual snowfall plotted on a graph since 1879, there is a modest decline for the first 50 years, and then no trend after that. The peak years for snowpack both occurred in the 19th century: 1880 and 1890 both had a maximum snowpack depth of more than 30 feet. The annual snowfall maximum (as opposed to snowpack) within the record was 1938 when 69 feet of snow fell at Donner Summit.

Despite minor fluctuations year to year, the Sierra Nevada remains a veritable snow factory, and will continue to be one for the foreseeable future. Surprisingly, just about all of the tremendously fertile Central Valley, one of the most productive agricultural regions on Earth, was a desert, until these great heaps of nearby snow were used for irrigation. According to Bar Yohai Davidoff of the California Department of Water Resources, agricultural water efficiency in the state stood at 72 percent in 1999 and in 2009 had risen to 74 percent. Although theoretically 100 percent efficiency is possible, it is not desirable because of negative consequences on soil composition, Davidoff says. The maximum efficiency for agricultural water use that preserves healthy soil is 85 percent.

Even using relatively inefficient irrigation systems during the drought between 1976 and 1978, agricultural production in the state of California *barely declined*. Since that time, improved irrigation techniques have become used increasingly. After conversion to drip systems, vineyards typically need just 20 percent of the water they used by flood irrigation, and vineyards have nearly all been converted. (Drip irrigation is not ideal for every crop nor for every soil.) In a state that has yet to attain its maximum water efficiency, net agricultural production has continued to rise. Just to give an idea of the state's extraordinary productivity, Fresno County, about 100 miles south of Yosemite Valley in the Sierras, produces 60 percent of all the raisins consumed *on Earth* and more than 90 percent of all the raisins consumed in the United States. Garlic, artichokes, strawberries, lettuce, citrus, avocados, and nuts are other crops in which California is not just a national leader but a world leader. And not one of them has shown a climate-induced decline.

Where is the trend that prompted Secretary Chu's remarks? Professor Daniel Sumner, of the University of California-Davis's Agricultural and Resource Economics School, confirms that water scarcity has not influenced crop production. "Yields have done nothing but go up." Here is a partial list of the agricultural products that have increased in tonnage and market price during recent decades: dairy (the largest cash producer in the state), lettuce, almonds, avocados, grapes, strawberries, and kiwifruit. Asked how the numbers on production jibe with Secretary Chu's assessment that there will be "a scenario where there's no more agriculture in California," Sumner takes a different view. "That itself is just ignorant," he says.

What about the threat posed by infectious diseases made more widespread by global warming? Here is an excerpt from a 2007 Bloomberg News story: "Global warming will put millions more people at risk of malaria and dengue fever, according to a United Nations report that calls for an urgent review of the health dangers posed by climate change." Stories like this one have been strewn all over the news for at least fifteen years, and they are based on appallingly bad fact-checking and even worse science. After watching distortion after distortion by the United Nations, one of the world's great experts on malaria and other insect-borne diseases, Paul Reiter, found himself forced, by his conscience, to withdraw from the IPCC. Although he resigned, formally, the IPCC listed him on its rolls, explaining that prior to his resignation he had reviewed some materials and thus his name belonged on the final publication. Reiter explained that he rejected the findings in the report and insisted on his name being removed. His request was denied until he threatened to sue.

Why was Reiter so offended in the first place? It turns out, for one thing, that *malaria is not a tropical disease.* It existed, to more or less devastating extents in most industrialized countries. Until the last century, when it was eradicated through pesticide use, it was a quite serious problem in the United States, particularly in the Southeast. Indeed the prevalence of malaria there is the reason that the Centers for Disease Control were put in Atlanta.

Reiter works for the Pasteur Institute in Paris and is routinely horrified to this day by the widespread misinformation regarding climate change and malaria. He brought to the public's attention that one of the worst outbreaks of malaria anywhere on Earth took place in the Soviet Union in the 1920s. In that single outbreak, during a time that the fledgling Soviet Union was struggling to hold itself together, there were 13 million cases a year and 600,000 deaths. The city of Archangel, near the Arctic Circle, saw 30,000 cases of malaria and 10,000 deaths.

So why, then, would study after study be published warning that malarial mosquitoes may increase their habitat by a few hundred miles north due to unprecedented global warming? The simple truth is that DDT nearly eradicated malaria in the early 1960s before the pesticide was singled out for negative attention in Rachel Carson's book *Silent Spring.* New efforts to use DDT on inside walls of homes are afoot, with some positive results. However, the disease currently kills more than a million people every year, mostly in Africa, a high proportion of them children, and millions more are horribly sickened.

Gore claims in *An Inconvenient Truth* that Nairobi, Kenya, was established in a safe zone "above the mosquito line," and that the city is now infested with mosquitoes because of global warming. This moved an appalled Reiter to publish an Op-Ed in the *International Herald Tribune*. "Gore's claim is deceitful on four counts," he writes. "Nairobi was dangerously infested when it was founded. It was founded for a railway, not for health reasons. It is now fairly clear of malaria. And it has not become warmer."

And yet, whose view is accepted as gospel by a majority of educated people in the West? Amazingly, it's not Reiter's. CNN, *The New York Times*, *The Washington Post*, and the BBC continue to toe the party line on malaria – that it is a "hot" disease made worse by anthropogenic global warming, rather than a symptom of poor public health policy and poverty. As long as these media giants promulgate global warming alarmism, esteemed scientists such as Professor Reiter are liable to stay on the margins of the debate.

A final pair of thoughts on malaria before moving on. In 1999, the British newspaper *The Guardian* published an article. "A report last week claimed that within a decade, the disease will be common again on the Spanish coast," the article read. "The effects of global warming are coming home to roost in the developed world." Well, not quite ten years have passed, and the Spanish coast is as malaria-free as it was a decade ago. Conversely, Shakespeare, who was familiar with malaria and its effects from seeing it close at hand in the England of Queen Elizabeth, wrote of it several times in his plays, a fact observed by Reiter as well. To repeat, malaria is not a tropical disease, and its reach has not been increased by global warming.

Yearning for a colder
past is one of the stranger
parts of the global
warming narrative.
Dramatic fluctuations
in glaciers and sea ice
have always been part of
the current Ice Age. *Ross
Ice Shelf, southernmost
navigable point on Earth.*

4 The Vanishing Ice Caps

The entire north polar ice cap may well be completely gone in five years. How can we comprehend the three million years, the period of time in which it has existed, and five years, the period of time during which it is expected to now disappear?

– Al Gore, December 13, 2008

One part of the global-warming narrative that isn't true is that the ice caps are in the process of catastrophic melt. They are not. One example: Greenland has areas that are losing ice, and others that are *gaining* it. The same is true of Antarctica: some gain, some loss of ice.

Misunderstandings about the region encircling the North Pole, specifically, and there are several in Al Gore's statement, are a long-running thread in the fabric of western culture. People who have never visited the Arctic, never studied its climate history, and who, if pushed, will admit that they know next to nothing, really, about the region, *believe* that they know a great deal about it. That is why, if you to dare mention that you have some doubts about the master narrative of global warming, the first question that you will hear is this: "What about the Arctic?"

Well, what about it? Fascination with Earth's

northernmost realm started at least as early as ancient Greece, as Robert McGhee sets forth in his book *The Last Imaginary Place:*

> The word Arctic comes to our modern languages from that of the ancient Greeks. It derives from "Arktos," the Great Bear, the constellation that we call the Big Dipper, which circles the northern sky without setting. This celestial phenomenon was thought of as so uncanny that the land that lay beneath it might also be a place where the ordinary laws of nature and society would not hold.

There is no shortage of entertaining ideas that people from southern climes have had about the Arctic. Here are a few: that it was an Edenic realm with ever-ripe fruit protected by tall mountains down which, toward known civilization, raged "Boreas, the bitter north wind that blighted northern regions and occasionally made life miserable even in the Greek archipelago"; that it was the "Land of the Amazons, another myth of antiquity"; that, if not Edenic, it was at least "mild and inhabitable"; that it was the gateway to the Earth's hollow, inhabitable interior from which emanated "luminous gases that occasionally escaped through polar orifices to cause the aurora."

Germans, in particular, McGhee writes, had a great knack for fantasizing about what was at the North Pole, inspired by the scribblings of Nietzsche. It was Germans, in 1912, who came to found the Thule Society.

Captains have been carefully navigating Arctic waters for at least 400 years, as the trade between the English and the Russians in the 1550s attests. Without the benefit of powerful modern machinery and multiple propellers, however, the historical ships' crews were out of options when the ice began to reform in the fall. *A NOAA vessels performs research in the Arctic Ocean, north of western Russia, 2006.*

This mystical German order believed that the island of Thule had been a northern Atlantis, home to an advanced civilization that was destroyed by catastrophe.... The Thule Society served as a recruiting centre for Bavarian radicals, who in turn formed the nucleus for the National Socialist Party organized in the early 1920s by Adolf Hitler.

As for the effort to create a blue-eyed master race with roots in the far north it was doomed from the start. Of the many peoples living around the Arctic Ocean's periphery during the last 11,000 years, not one has been known for such coloring.

Alongside the fantasizing and myth-making, McGhee demonstrates, a now millennia-long pattern of northern exploration also got its start during ancient times. After transiting the Straits of Gibraltar, in the fifth century, B.C., the Greek mathematician Pytheas sailed at least as far north as the Scottish island of Shetland, and possibly all the way to Norway. His tales of mystical gray scenes were likely attempts to describe sea ice, fog, snow, or all three.

Thirteen hundred years after Pytheas's trip, Vikings departing from Norway settled Iceland and then Greenland, keeping a colony in the latter for 350 years before the cold returned. The first Europeans to explore the southern reaches of the Arctic Ocean profitably were the British. They had been eclipsed in less risky efforts at empire building by the Spanish and the Portuguese, whose fleets had formed outposts and eventually colonies in the relatively forgiving environment of the Caribbean. Britain had become convinced that a navigable Northwest Passage or Northeast Passage had to exist, making efforts at forming trade partnerships with anyone living on or near the Arctic coast a priority. Although the ice prevented the English from sailing all the way from the top of Norway to the Bering Strait, Captain Richard Chancellor did get his ship to an important Siberian river mouth.

That voyage led to many years of important trading between the Russians, who provided furs, and the English, who provided woolen textiles. Half a century before landing at Plymouth Rock, thus, the British were carefully plying the seasonally ice-choked waters of the Arctic.

The ocean's annual and decade-long accordion-like expansions and contractions of sea ice were among the reasons that mariners imagined that there *must* be a route all the way through to the other side, via the Northwest Passage (across the top of Canada) or the Northeast Passage (across the top of Siberia). The only thing that was stable about the ice, it turned out, was that it changed from year to year. Many an expedition seeking the Northwest Passage was lost, as suddenly forming ice in the fall bound ships for the duration of the winter. Several search parties were then lost themselves pursuing those that had gone before. The most famous of the lost expeditions, said to have ended in cannibalism, was one led by the accomplished explorer and mapmaker John Franklin. A *series* of rescue missions were launched to locate Franklin's lost ships, and a goodly number of these trips ended in disaster. To our eyes, sending men into the jaws of death in this way can appear to be a form of madness, but the ice was known to have shifted, dramatically, over the years, decades, and centuries. This fact, coupled with the belief that the Franklin party had progressed a long way toward finding the Northwest Passage was all the motivation that the rescue parties, and their sponsors, required.

Not all of the Franklin-rescuing trips ended in tragedy. The best-known of the many individual ships, British and American, to seek the fate of, or potentially rescue, the Franklin expedition was the *Resolute*, captained by Henry Kellett. Part of a four-ship flotilla led by Edward Belcher, Kellett's ship became frozen in sea ice deep in the Canadian Archipelago in the fall of 1852, remaining trapped for two winters. (One good history of the expedition is Martin W. Sandler's *Resolute*.) When the ice showed no sign of releasing his ship, Kellett led his men on sleds to another of the expedition's ships that was bound fast. Captain Belcher, in the end, decided that all hands should be transferred to the only ship that had managed to find open water. The decision to abandon *Resolute* led to a court martial in which both captains were acquitted.

As it turned out, moving pack ice carried the abandoned ship 1,200 miles, from Dealy Island down into Davis Strait, between Baffin Island and Greenland. It was there that the crew of an American whaler, *George Henry*, noted it. The Americans were able to free *Resolute*, re-rig it, and sail it to New London, Connecticut. Although the British waived all rights to the ship, an American merchant, Henry Grinnell, convinced the U.S. government to restore *Resolute* to immaculate condition and sail it back to England as a friendship gesture. The ship was presented to Queen Victoria at a ceremony held in Cowes harbor on the Island of Wight on December 17, 1856. A couple of decades later, the British government had a desk made from the timbers of the by-then decommissioned *Resolute* and presented that desk in 1880 to President Rutherford Hayes. It has been the principal desk used by U.S. presidents in the Oval Office ever since.

A Northwest Passage was eventually transited, by the world-famous Norwegian explorer Roald Amundsen in 1906. The Swede Adolf Palander had completed the Northeast Passage a few decades earlier, moving west to east. (The North Pole, an object of fascination of its own, was attained by dogsled, probably first by the American explorer Robert Peary in 1909.)

Obsessive interest in the North Pole and the Arctic Ocean with regard to global warming got going in earnest in the 1990s. Satellite measurements of the Arctic sea ice had begun in 1979, and an ominous downward trend was becoming visible. However, if it had been possible to fly the satellite in 1945, the record would have shown a 30-year increase in ice at the same time that carbon-dioxide emissions were increasing. Indeed, thickening and growth of Arctic sea ice as measured by the U.S. Navy's fleet of nuclear submarines were among the reasons for fears of a coming Ice Age in the late 1960s and early 1970s.

And if the Arctic explorers seeking a Northwest Passage and the North Pole could have traveled back in time to the Holocene Climactic Optimum, they would have had an easier go of it. That is because there was substantially less ice in the Arctic at the time. Scientists have discovered ridges in the sand along Greenland's northern shore left by wave action 6,000-7,000 years ago.

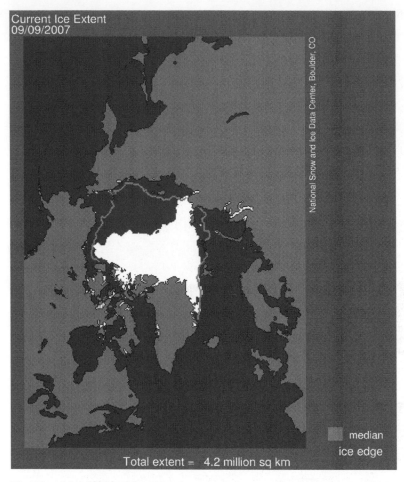

Current Ice Extent
09/09/2007

National Snow and Ice Data Center, Boulder, CO

median
ice edge

Total extent = 4.2 million sq km

The idea that recent melting is a first-time-ever event is contradicted by the discovery of wave patterns in the sand left on Greenland's north coast when at least some of the Arctic was open water 6,000-7,000 years ago. *The 2007 melt, shown within the anomaly line derived from 30 years of satellite data.*

As for the idea that what is occurring in the current decade is unprecedented in the last couple of centuries, a detailed analysis of Arctic explorers' logbooks during the 19th and early 20th centuries suggests otherwise. Scientists, including J. L. McKay, have corroborated the fact that sea ice has changed throughout the Holocene interglacial. As McKay wrote in 2010, "The results of this study clearly show that sea-ice cover in the western Arctic Ocean has varied throughout the Holocene. More importantly, there have been times when sea-ice cover was less extensive than at the end of the 20th century."

Although scientists are aware of the Arctic's physical characteristics and history of massive climate change, Al Gore does not share that awareness, evidently. When the former vice-president stated in 2008 that the Arctic Ocean would be ice-free within five years, it was evidently under an assumption that the decline in sea ice shown by the satellite from 1979 to 2007 would continue, and even accelerate, as it had at the end of the period. But such an assumption ignores the significant warming and melting of the 1930s and '40s, and ignores, too, the significant expansions and contractions of the ice during the previous centuries, and millennia.

This brings up an important point: Gore's statement that the Arctic sea ice had been there continuously for three million years is simply false. Even at the end of the last 100,000-year period of glacial advance, the oldest ice at the North Pole was likely no more than about 100,000 years old. That is because the great majority of Arctic Ocean sea ice *always melts during interglacials*. The interglacial before our own, the Eemian, which lasted from 130,000 years ago until 110,000 years ago, was much warmer than ours.

Indeed, the Eemian was so much warmer than the Holocene that sea levels were somewhere between thirteen and nineteen feet higher than today. This higher sea level had nothing to do with the ice in the Arctic, however, as that ice is *floating on the surface of the ocean*. If all of it were to mysteriously melt overnight, tomorrow morning the world sea level would be unchanged. The same holds true for the ice floating at the top of a glass of water; after the ice melts, the water level stays the same. In the Arctic, the higher sea level during the Eemian came from glaciers, and not from locally melted sea ice.

So, during the warm Eemian interglacial the Arctic Ocean sea ice melted to nothing, or next to it, although the event had no effect on sea level. Scientists believe that this is standard: The Arctic Ocean has little or no sea ice older than a human generation during interglacials, including the Holocene. Old ice cannot accumulate at the North Pole, because the pack ice is in a constantly-moving train, on its way east and south toward the North Atlantic, where it eventually melts out. Calling the Arctic sea ice an "ice cap" during an interglacial is somewhat misleading. Unlike the "hard hat" of land-bound glaciers in Antarctica, the Arctic sea ice is like a knit garment that is rapidly stitched together on its edges each fall and slowly pulled across the Arctic basin toward the relatively warm waters off southern Greenland. When the transited ice completes this trip, the southernmost portion of it melts, principally in late summer. It can and does take some parts of the fabric as long as a few decades to make the trip, but, again, it is a highly variable amalgam of ice, some brand new and some older, that sits atop the Arctic Ocean each winter.

For much of the year, there are holes in the Arctic sea ice, as in a knit garment. These open spaces are called "leads." They come and go with the variation of current, and the massive forces at work within the ice. The leads create an optimal environment for polar bears, who rely on them to hunt seals that come up for air.

Contrary to popular belief, polar bear populations have not lowered during the past 30 years. On the contrary, the bears' numbers have rebounded. In terms of changing sea ice patterns, the bears are equipped with the ability to swim as far as 60 miles in the open ocean, even if they seldom are required to prove themselves in such a way. Popular notions of their drowning when separated from land by melting ice are not based on science. The animals came through the Holocene Optimum, and Medieval Warm Period, and the more recent warming during the 1930s and '40s in fine shape.

Polar bears are, however, threatened by hunting, especially from aircraft. Commercial polar-bear hunts became popular in the 1960s. After bottoming out in the early 1970s somewhere in the vicinity of 5,000 bears (when the Arctic was significantly colder than in 2009), bear numbers are back up to somewhere between 20,000 and 25,000, primarily because hunting has been curtailed. In the early twenty-first century, the Canadian government continues to give out permits for the killing of 800 bears per year, but there is *plenty* of ice, plenty of cold, and plenty of seals in the Arctic basin to support polar bear populations at their current levels. What the

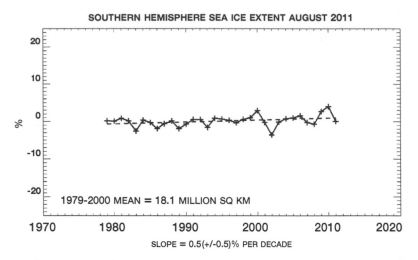

So far, Antarctica's sea ice has increased during the satellite era. *Graph from National Snow and Ice Data Center, comparing ice extent during month of August.*

white-coated bears simply cannot tolerate are bullets.

What about the telltale year of 2007? Although, like the Arctic, Antarctica has only had its sea ice measured by satellite since 1979, its story is compellingly different. Since the satellites have been flown, Antarctic sea ice has steadily *increased*. And, once again, the record year for Antarctic sea ice? 2007.

Interestingly, though, when images are shown of glacial calving, generally speaking they are photographed in Antarctica – the land of *growing* sea ice. How can this apparent paradox be resolved? Simple: Glaciers that are exposed to sea currents on their periphery, such as those on the Antarctic Peninsula (jutting up toward the southern tip of South America), have been calving just as long as they have existed. These photogenic incidences of glacial calving (the big splash-downs of ice into liquid sea water) happen at the boundary of sea ice and land-based glaciers. Of course, Antarctic sea ice is a non-static system and can never be static. Liquid water rushing against frozen water will, eventually, cause parts of the ice to melt. It is well known, too, that some glacial calving in Antarctica is caused by the growth of land-bound glaciers pushing ice down to the sea.

Meanwhile, both polar regions have witnessed cyclical glacial calving throughout the current Ice Age, which began about three million years ago. *Much* more spectacular glacial calving has taken place at different points during the Holocene interglacial as well as during other interglacials than anything that has been videotaped by people bent on "raising awareness" during our time. Scientists know this from the high-water stands measured during interglacials (such as the Eemian), during the Holocene Optimum 9,000 years ago to 5,000 years ago, and during the Medieval Warm Period from roughly the year 1000 to the year 1350. What did not exist, during any of these ancient times, when the chunks of ice tumbled, was a technology to record the allegedly "disastrous warming" in the form of video cameras.

Antarctica, unlike the Arctic, is a continent, and its miles-thick ice sheet sits atop dry land. Antarctica's sea ice, unlike the beanie up north, is a skirt around this land. And the scale of things down south is bigger. The skirt of ice, being farther from the pole than in the Arctic, melts nearly 100 percent each summer. When it expands in the fall and winter to its greatest extent, it occupies more area than the Arctic sea ice does. And the riddle remains: In 2011, Antarctic sea ice is more abundant than in 1979.

If, as the United Nations scientists have presented, carbon dioxide is a well-mixed gas in Earth's atmosphere; and if that gas has the effect on atmospheric temperature that the UN says it has; and if that effect is magnified at the poles as the UN says it is, causing, as the story goes, the melting of the Arctic sea ice; then something is terribly rotten in Antarctica. In a somewhat desperate gambit, NASA, the UN, and various university scientists have argued that the additional sea ice surrounding Antarctica is not indicative of continent-wide cooling – that there is no paradox here. What they say is that the ozone hole over Antarctica has facilitated the growth of sea ice around the southern continent, through a process that involves the strengthening of the winds ringing the continent's perimeter. Most of these same scientists argue that, as a whole, Antarctica is warming, if only slightly.

But let's stick with the facts: Antarctic sea ice is *increasing*, and has been during the entire period of satellite measurement. Here's another fact: the temperature at the South Pole is *falling*.

It turns out, though, that getting reliable measurements from automated temperature stations across the rest of the continent is quite difficult, as the stations routinely get buried in snow and have to be dug out. While covered, their readings are falsified by snow's insulating property. (It is a little-known fact that there are scientists whose salaries are paid by various national governments to check on the effect of global warming in Antarctica who spend significant percentages of their time unburying temperature stations from accumulated snow.)

What of the Wilkins ice shelf and other named ice shelves, which photographers and videographers love to show shattering along the Antarctic Peninsula? Don't *these* prove the existence of manmade global warming? No, they do not, and here is why. Ice shelves are distended glaciers with some of the characteristics of land ice and some of the characteristics of sea ice. Seated on both underwater rock and land that is above the water line at the immediate coast they extend many kilometers away from land, becoming increasingly tenuous in the process. At their outermost edges they are effectively sea ice, as can be easily seen when they break up, with chunks of the shelves floating right next to the sheets that they were part of the day before.

The melting of these shelves is not a recent phenomenon. They used to extend hundreds of kilometers farther out from the coast than they do today. They have been melting, in fits and starts, throughout the Holocene interglacial, as they do during every interglacial. What mostly causes the buckling and collapse of ice shelves is not warm air above them, but relatively warm currents from below.

The idea that the famously fragile Wilkins ice shelf existed during the warm Eemian interglacial is preposterous. That segment of ocean has had ice on it, and no ice, in succession, dozens of times. Bear in mind, too, while we're on the subject of Antarctica, that it was once inhabited by dinosaurs who lived in a leafy tropical environment right there at the bottom of the Earth. Of course, things have cooled down, globally, quite a bit since then.

One last thing, before leaving the subject of the ice caps. In January of 2009, *Nature*, long one of the most highly respected scientific journals, arguably embarrassed itself by publishing a cover article with the headline "ANTARCTIC WARMING." The article's lead author was Eric Steig, but it was based on a statistical ploy by a famous climatologist named Michael Mann. Mann nearly, but not quite, deserves a chapter of his own, for so skillfully skewing the public discussion of climate. He famously did this with a single graph that erased the Medieval Warm Period *and* the Little Ice Age and indicated that the last half of the 20th century was the warmest period during the last 1,000 years (by an alarming amount). After triumphant use by the IPCC and Al Gore at the start of the current decade, the graph was debunked by a U.S. Congressional committee and has not been used in subsequent IPCC reports and conferences.

For the *Nature* article on Antarctica, Steig and Mann devised a statistical means of indicating warming more or less throughout Antarctica. In a nutshell, they took some mild warming from western Antarctica, especially the Peninsula, and generalized it eastward, covering the entire continent. But 98 percent of the continent remains bitterly cold, and no one disputes that. Also, the temperature recorded at the South Pole is agreed to have declined during the last 75 years. Finally, it has been suggested by competitor researchers that some of the data Steig used came from buried stations, which give a warm bias. With or without the snow-buried stations, the Steig article in *Nature*, and the red-tinged image of Antarctica on the cover of the magazine, give a profoundly misleading sense of whatever is taking place at the bottom of the Earth, where sea ice is growing and the South Pole is cooling.

Volcanism, a further complicating factor, has been shown to be on the rise in west Antarctica as well. The extent to which geothermal heat from the scores of volcanoes in the region may have contaminated Steig's data is not known, but it is not a subject that he and his team explored. Further, by carefully choosing his start and end dates for the study, overusing data from the warming west, downplaying data from the cooling east, and ignoring the issue of volcanic heat, Steig was able to show a net "continent-wide" warming of 0.1 degree Celsius per decade, or half a degree in 50 years.

Even with Steig's skewed data, it would take *thousands of years* for the ice sheet to melt at such a rate.

As I write this, the start of winter in the Southern Hemisphere is three weeks away. The temperature at Vostok, Antarctica is -84° Fahrenheit. At the same time, residents of coastal Florida are told to feel anxiety over the incipient flooding from the melting of Antarctica.

Among the many world-class scientists who take

exception to the view that Earth's ice caps are imperiled are Duncan Wingham and Cliff Ollier. Wingham published an article in the journal of the Royal Society showing that global warming is definitively *not* occurring on Earth's southernmost continent. Canadian journalist Lawrence Solomon wrote of Wingham and his co-authors:

> By studying satellite data from 1992 to 2003 that surveyed 85% of the East Antarctic ice sheet and 51% of the West Antarctic ice sheet (72% of the ice sheet covering the entire land mass), they concluded that the Antarctic ice sheet is growing at the rate of 5 millimeters per year (plus or minus 1 mm per year). That makes Antarctica a sink, not a source, of ocean water. According to their best estimates at the time, Antarctica will "lower global sea levels by 0.08 mm" per year.

Wingham's observations are confirmed, nicely, by the buried temperature stations.

The other region pertinent to any discussion of the polar ice caps is Greenland. Hundreds of articles and broadcasts have mentioned its ice as a source of the water that will raise the world sea level. Some of the journalists (and some of the scientists) who produced these articles would have done well to check with the Australian geologist Cliff Ollier, a specialist in glacier and ice sheet behavior. For Ollier, hearing mainstream media scare stories about Greenland and Antarctica is, simply put, torture. That is because they repeat the same inaccuracy,

Thirst for knowledge of both poles and their environments has been ever greater in the west for the last several centuries. *The Terra Nova, the ship used by the doomed members of the British Antarctic Expedition, at McMurdo Sound, Antarctica, circa 1911.*

again and again. The usual claim by computer modelers, and the journalists who treat their findings as actual data, is that the ice sheets are on inclined planes made slippery by water melting beneath them and that the entire system is unstable, and increasingly so. In fact, Ollier has stated, "The Greenland, East Antarctica, and West Antarctica ice sheets occupy kilometer-deep basins, and the ice cannot possibly slide downhill. After three-quarters of a million years of documented continuous accumulation, how can we believe that right now the world's ice sheets are collapsing?"

In addition to the plight of the (thriving, it turns out) polar bears, one other play at the heart strings is a standard part of our modern-day myth-making regarding the Arctic. That is this: The native peoples of the far north were living in Edenic, if frosty, terrain until the warming of the last century.

In fact, the native peoples of the north have had to contend throughout the last 10,000 years with intense climatic variation. Although temperatures from 8,000 years ago to 6,000 years ago were warmer than today, and land in northern Siberia that is now tundra was heavily forested, the various Arctic peoples were not forced from the region that they call home. After the cold times of the Dark Ages, the Arctic was again affected by global warming when the Vikings and the Inuit co-existed in Greenland during the Medieval Warm Period. Did the Inuit, for instance, perish during this time of relative Arctic warmth? They did not. Did they find it stressful? Yes, they probably did, as Robert McGhee notes:

To Inuit hunters who seem to have been adapted primarily to hunting on land and sea-ice, the unexpected appearance of open water during early summer or a delay in the expected freezing of autumn seas would have brought hardship and often disaster. Inuit today worry that the warming seen over the past decade will bring disastrous changes to their way of life, and the situation a thousand years ago would have been as serious.

So, although sea ice was almost certainly at least as sparse as now, the Inuit found prey enough to survive. Later, intense cold returned for centuries during the Little Ice Age, whose last vestiges have come to be viewed by modern humanity as "normal."

Toward the conclusion of the warming of the 1930s

and '40s, a famous trip through the Canadian Archipelago took place. This was the sailing of the *St. Roch*, the first known vessel to traverse the Northwest Passage in a single season. Bearing out the warm conditions making such a trip possible, numerous weather stations from the Arctic show annual mean temperatures during the 1930s and early 1940s warmer than today.

Following the melt cycle of the 1940s, came a descent of the ocean-atmosphere system into newfound cold, with Arctic sea ice growing to its maximum extent at precisely the same moment that human beings happened to get a satellite into space to begin tracking it. The thirty-year period of sea-ice decline that has taken place recently, so ballyhooed by Al Gore and various journalists (and, less explicably, by dozens of prominent scientists), has every likelihood of being no more significant than any other natural cycle.

Computer models have come to replace reality in the public debate about climate. *NASA's "Columbia" supercomputer, Mountain View, California, 2006.*

5 Rise of the Machines

Like one that draws the model of a house
Beyond his power to build it; who, half through,
Gives o'er and leaves his part-created cost
A naked subject to the weeping clouds
And waste for churlish winter's tyranny.

– William Shakespeare, Henry IV, Part II

So the ice caps aren't melting. That doesn't mean there isn't plenty of other ammunition with which to scare otherwise sane people badly. Put yourself in the shoes of the global warming doomsayers. If you're going to scare the pants off a whole bunch of folks, you're going to need some powerful tools. Arguably the most powerful tool available for such a purpose is the supercomputer. That is why a single phrase appears in nearly every article and book dealing with climate change. Although the phrase is used in other disciplines to signify divergence from reality, in the case of climate science it has come to be equated with reality, or even to replace reality. Its proponents are passionate, tireless. Its detractors don't really know where to start.

That phrase? *Computer models.*

The full formal name is **general circulation models**. Using gridded **cells** the size of Connecticut, these computer models are humanity's effort to lasso, intellectually, what may be the most mentally uncontrollable being ever created: Earth's climate system. There are many, many issues with models.

Using the most powerful supercomputers in existence, modelers strain to generate even faintly accurate climate forecasts, simply for lack of computing power. *The ocean-atmosphere system is that complicated.* Among the items that the models must attempt to compute: highly complex, poorly understood deep-sea currents; the effects of **aerosols** (fine pollution particles) on **cloud formation**; the effect of **black carbon** pollution on the melt rate of snow and ice, especially in the Arctic; **solar radiation** (via an effect known as **solar dimming**); volcanic eruptions; the effect of air masses of different pressure on either side of mountains (a process known as **mountain torque**); variations in wind patterns, particularly of trade winds that lead to El Niños and La Niñas; variations in **albedo,** which is the extent to which the Earth's surface and atmosphere (ice sheets, clouds, oceans, forests, deserts, cities, farms, rivers, and lakes) reflect radiation back to space; and, finally, **solar variation,** including a controversial secondary effect of the Sun's shifting phases on cloud formation in our atmosphere.

Every one of these variable quantities is being debated in the scientific literature. And in the blogosphere the debate is red hot, as exemplified by the tongue-in-cheek suggestion by more than one blogger that global warming skeptics be murdered in their sleep. Controversy aside, just *measuring* any one of the factors to be included in computer model simulations, at any given moment in time, is nearly impossible. Among the reasons: The planet is a lot bigger than the average person gives it credit for being. The ability to fly from one continent to another in less than half a day gives a false impression of scale, it turns out.

My adoptive home state of Texas can help convey the size of the planet. Texas, at 268,581 square miles constitutes 13.2 percent of the land area of the United States (which comes in at 3,537,441 square miles). The surface area of Earth, though, is 196,935,000 square miles. The percent of the world's surface occupied by Texas, then, is 0.5. And yet within this one-half of one percent of Earth's surface area are several vastly different climates. From the high mountain desert of El Paso in the west, where accumulating snowstorms are typical most winters, to the southernmost coast on South Padre Island where the warm Gulf of Mexico water acts as a powerful buffer against temperature extremes (especially those of winter), to oppressively hot and humid Houston, to Amarillo in the Panhandle with its four distinct seasons, to the state's myriad river systems (each with its own micro-climate), Texas (as those who have labored to drive across any portion of it know) is enormous. Likewise, every other half-percent of the globe's land mass is, too.

All of this is to say that size is a problem for the computer models. They have to deal with something that is giant. Breaking it down into cells the size of Connecticut makes sense in the face of a project of such scale, and yet even little Connecticut has myriad micro-climates within it. Snow on the coast and none in the Berkshires, and vice-versa, are commonplace in the state during winter. Ninety degrees in Hartford and 57 degrees in Greenwich on a late-spring day with a good sea breeze is not uncommon, either.

While the list of issues with which computer models must contend is long, nothing is so confounding for those who would predict our climate's future anywhere from thirty to a hundred years from now as the next two items: **clouds** and **tipping points**. Although believers in manmade global warming will tell you otherwise, the physical processes involved with atmospheric dynamics remain, to a shocking extent, theoretical. As an example, no model being used by NASA or any of its peer-organizations has any viable allowance for bacteria in the formation of clouds, as the research into the phenomenon is only getting started. And, yet, it turns out, this is just one of several aspects of cloud formation that remain poorly understood. As for tipping points, one of the most frequently referenced is the Arctic ice "cap." While it has been posited that removing the ice from the Arctic basin would increase the planet's temperature dangerously, this has never been *shown* to be the case. And yet this supposed tipping point is factored into the majority of computer models as though it were an objectively measured fact. It is not. It is possible that the opposite is true: That if you wanted to *cool* the planet suddenly, one of the best ways to do so would be to remove the insulating ice covering the Arctic Ocean.

That is because heat from the tropics is constantly being transported northward. The vertical Sun strikes the ocean in the tropics, warms the water, and currents carry the warm water toward the North Pole. It is largely this warmth of tropical origin that makes the temperature stations around the Arctic Ocean warmer in winter than those farther south, in Siberia for instance, away from the maritime influence. As an example, a typical wintertime temperature at noon in Verkhoyansk, Russia, is -50 degrees Fahrenheit, whereas the temperature at the North Pole at the same time on the same day is frequently 25 degrees Fahrenheit *warmer.*

This exemplifies a single area of inquiry into the ocean-atmosphere system that remains more or less mysterious. The study of Earth's climate is, after all, in its infancy. The removal of the Arctic ice cap is a theoretical tipping point, however, among several others that the scientists running global circulation models present as *gospel.* Another prominent tipping point is that of the Greenland ice sheet. The theory goes that if the atmosphere over Greenland warms a few degrees Celsius, then the ice sheet could more or less suddenly melt. This particular tipping point requires an understanding of the ice sheet as inherently fragile. But we know that this hasn't happened during three-quarters of a million years of steady thickening of the ice – during warmings far more notable than what is being observed today. Nonetheless, the melting of Greenland's ice sheet, too, is among the most feared tipping points, theoretically leading to the submersion of much of Bangladesh, the Netherlands, Florida, and Manhattan.

Which is to be accepted as authoritative: a model or reality? An *imagined* tipping point, or an observed stasis? Tipping points are the most dramatic sub-set of a larger group of factors that computer models must include. That larger group is **feedbacks**. There are at least a dozen significant feedbacks within the Earth's ocean-atmosphere system, and many of them influence one another. This can be seen in the example of the trade winds. Trade winds in the Pacific Ocean govern El Niños and La Niñas. Easterly winds moving from the coast of South America toward the Philippines, the trades slacken and strengthen cyclically, and are themselves driven by a number of forces: solar variation, variations in cloudiness where the trade winds originate, ozone variation and ozone transport, fluctuating cells of high pressure off the California and Chilean coasts, the warm water from a building El Niño itself slows the trade winds (leading to a feedback mechanism that can intensify that El Niño). Once you have either an El Niño or a La Niña, you get cascading feedbacks throughout the global circulation: abundant precipitation in many locations, drought in others, changed wind patterns, and greater or lesser release of CO_2 by the oceans. Indeed, looking at the most widely respected record of CO_2 variability, the one generated by the National Office of Atmospheric Administration on Mauna Loa on the Big Island of Hawaii, unusually cold years associated with La Niñas are correlated with smaller increases in CO_2. Warm years associated with El Niños are correlated with larger increases in CO_2. This occurs through the process known as **degassing**, the release of dissolved CO_2 from water. At certain ocean locations where this occurs most, the seawater literally fizzes.

Theoretically, this could lead to a bad situation, whereby increased warmth in the Equatorial Pacific leads to increased release of CO_2, which leads to warmer atmospheric temperatures, which then lead to warmer Equatorial Pacific waters, and so on. It is something like this that the computer models predicted when they first started indicating in the late 1980s that temperatures would climb to unprecedented levels within a century. The first domino posited by the modelers in this progression was the increase in CO_2 from human activity.

The idea that rising CO_2 would drive the ocean-atmosphere system off the rails is clear in the following comment by Russ Schnell, a scientist doing atmospheric research at Mauna Loa Observatory in Hawaii. At a 1997 conference in London, Schnell explained the risks of such a chain of feedbacks originating from the tailpipes of America's SUVs. "It appears that we have a very good case for suggesting that the El Niños are going to become more frequent," Schnell said, "and they're going to become more intense and in a few years, or a decade or so, we'll go into a permanent El Niño."

There is, it must be said, a certain logic behind Schnell's prediction. That logic draws heavily on the fact that the strongest El Niño ever measured directly was under way in the tropical Pacific. (In fact, the ocean-atmosphere system is still being influenced by the heat discharged by that El Niño twelve years later.) This is why it is possible to say that fourteen of the "hottest years ever" have all occurred in the last fifteen years – the great El Niño of 1997 pushed the Earth's mean temperature that far up the hill.

But there is no evidence that 1997's El Niño was produced in response to warmer atmospheric temperatures that preceded it, or to carbon dioxide. The amount of heat involved is vast, with even a weak El Niño, let alone one of such great scale. The idea that such heat could cross the air-water boundary via radiative transfer, even cumulatively over the course of many years, is a weak theoretical construct. And yet El Niños come into existence over the course of weeks or months. Although sunlight is known to reach a hundred meters or more into the upper layer of the ocean, radiation emitted by the atmosphere penetrates only a few millimeters. The truth is that not one of the modeling teams can reliably predict the emergence of an El Niño – in terms of initiation, scale, or duration – a year before it comes to pass. But to predict where the ocean-atmosphere system is going using a computer model, you need to be able to predict precisely that, only from a far greater distance than a single year.

It is not that the modelers haven't tried. James Hansen's Goddard Institute team has been predicting another Super El Niño, for several years, in the apparent hope that such an event would establish the accuracy of their global circulation model. However, other scientists identify the 1997 El Niño as a once-in-five-hundred-years natural phenomenon.

The computer modelers, on top of presuming that the 1997 El Niño was representative of an oppressively hot future, simply *presumed* that it was caused by forces external to the known ocean-atmosphere system. Since Schnell's prediction of increasingly frequent, and then ever-present, El Niño conditions 13 years ago, however, there have been four more El Niños, and there have been three La Niñas. Contrary to the prediction of ever-heating oceans and an ever-increasing global mean temperature, the La Niña of 2007 pulled the satellite-measured global mean temperature down to the *same level* where it had been when the satellites were put in orbit in 1979.

Remember, this analysis started with a discussion of trade winds. The progression goes like this: slackened easterly trade winds, increased **insolation** (more sunlight on the same patch of water), the sloshing back toward the east of the great warm pool that accumulates in the western Pacific when the trades are normal, greater release of CO_2, warmer atmosphere. But another important feedback for the computer modelers is this: water vapor.

Many computer modelers will allow that CO_2 has limits as to how much warming it can provide. They accept that its warming properties operate on a logarithmic curve, and that a doubling of carbon dioxide from present values will not itself yield a doubling of the warmth that the gas provides. Rather, the most widely circulated figures for how much warmth a doubling of atmospheric carbon dioxide would provide vary from 0.5° Celsius to 1.5 ° Celsius. What the modelers figure, though, is that when you warm the oceans, you will generate a great deal more water vapor in the atmosphere through evaporation. And this is true. They then *presume* that this vapor will act globally the same way higher humidity works locally on a warm summer night. The problem is that water vapor is implicated in a panoply of complex feedback mechanisms. The most complex of them is clouds.

If you knew how little scientists know about clouds, you would be astonished. An example: Galactic cosmic rays generated by the birth of stars deep in our own galaxy may play a prominent role in cloud formation. But leaving aside where they come from for a moment, what clouds do with regard to the transfer of energy is poorly understood. Low clouds reflect sunlight back to space and thus have a net cooling effect. High clouds, the thin milky ones formed by jet aircraft and those that are formed naturally as well, have a net warming effect. Towering storm clouds over the tropics are believed by most atmospheric physicists to reflect nearly all sunlight from their tops back to space, and, thus, have a net cooling effect. Another point: Clouds transport huge quantities of energy from Earth's surface upward toward space by the process of convection. Warm air rises, often violently. Convection is notoriously difficult to model, in and of itself.

In sum, because the science of cloud formation remains shrouded in uncertainty, *computer models cannot even begin to approximate the global effect of cloud cover.* Thus, the effort to assess the role of water vapor on the global mean temperature is rendered moot before it can properly begin. Nonetheless, the IPCC, using models, figures clouds to have a net warming effect of more than one degree Celsius.

Despite its aggressive attribution of warming to clouds, the IPCC's Fourth Assessment Report, published in 2007, openly discusses the limitations of models, as well as specific issues having to do with clouds:

> The ultimate source of most such errors is that many important small-scale processes cannot be represented explicitly in models, and so must be included in approximate form as they interact with larger-scale features. This is partly due to limitations in computing power, but also results from limitations in scientific understanding or in the availability of detailed observations of some physical processes. Significant uncertainties, in particular, are associated with the representation of clouds, and in the resulting cloud responses to climate change.

Uncertainties regarding clouds remain large enough to make it unclear whether the planet will warm by one degree Celsius, four or five degrees Celsius, or indeed at all, if carbon dioxide is doubled.

Like so many policymakers, United States "science czar" John Holdren is clearly concerned about the fate of humanity in the face of some of the scenarios generated by computer models. Associated Press science writer Seth Borenstein noted that the director seemed to have one thing on his mind during his first on-the-job interview in

The use of computers to anticipate the atmosphere's doings began in the 1940s. In 2011, getting a read on the atmosphere three days out remains enormously challenging. *Using a Weather Bureau computer, 1965.*

April 2009. "Twice in a half-hour interview, Holdren compared global warming to being 'in a car with bad brakes driving toward a cliff in the fog,'" Borenstein wrote.

But could not the director's statement just as easily be applied to retooling the world's economies in response to computer models?

We have not yet even considered two of the greatest challenges the modelers face: **parameterization** and

initialization. Parameterization is the process by which the computer is told what to consider within the *infinitely complex climate system:* deep ocean currents; surface currents; turbulence within all layers of the ocean; ocean heat content (the energy contained by the ocean in its top several hundred meters); ocean surface temperatures (measurements obtained from the top few meters of sea water); wind patterns; atmospheric pressure patterns; aerosol distribution; precipitation patterns; land-use data (tracking how much land is being cultivated, plowed under for urbanization, or left in natural state as desert, rainforest, prairie, tundra, temperate forest, alpine forest, marshland, etc.); glacial mass; ice sheet area and mass; sea ice extent and volume; soil moisture content; global mean relative humidity; global mean temperature; solar variation; rate of greenhouse gas emission by humanity; rate of greenhouse gas emission from natural sources; and historic responses of the ocean-atmosphere system to changes in any and all of these values on time scales varying from several years to several centuries.

So, that is parameterization. On a practical level, it is a script written by scientists which the computers must read. A computer cannot, for instance, assess the ocean-atmosphere system on its own. It only knows what scientists tell it, leading to the famous "garbage in, garbage out" saying on the part of model skeptics. (By the way, skeptics also like to point out that it was another network of computer models, in this case financial ones, that created the economic catastrophe of 2008.

As for initialization, the news is bad here, too. Picture all of the parameters named above and a couple of dozen others that could be added to the list. Understand that not one of these can be frozen in time; all are dynamic. They can be viewed as three-dimensional domino structures stretching as far as the eye can see in a vast warehouse – up, down, left, right, behind. Cascading patterns of dominoes falling without end. Yet, somehow, *climate modelers must stop the dominoes*, theoretically, in order to *initialize* the global circulation model that they would run.

Finally, the entire ocean-atmosphere system is non-linear. Another way of stating the same thing is that it is *chaotic*. Small events on one end of the system have enormous consequences on other parts of the system. This was made famous by Edward Lorenz, the father of chaos theory, in the form of the Butterfly Effect. The Butterfly Effect theory holds that the flapping of a butterfly's wings in Hong Kong is capable of producing a hurricane thousands of miles away. So, while some scientists have convinced themselves that adding a given amount of a trace atmospheric gas to the system is pushing a domino that leads to dire atmospheric results, and have sold such a linear view to the great mass of concerned non-experts, X (added CO_2) + Y (a stable, linear ocean-atmosphere system) = Z (a dangerously warmer system) in no way approaches reality. You cannot leave chaos theory behind when evaluating climate. Nor can you leave behind the reality of multiple negative feedbacks. If comforting facts such as these are not one's cup of tea, though, there are always computer models. Properly parameterized and initialized, they can scare an otherwise sane person to the core.

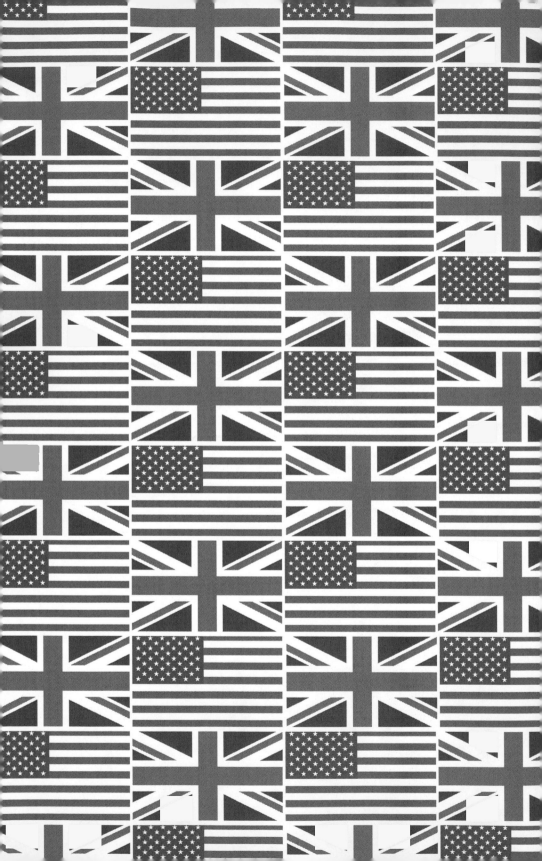

6 The Master Narrative

Our long-term security is threatened by a problem at least as dangerous as chemical, nuclear or biological weapons, or indeed international terrorism: human-induced climate change.

– Sir John Houghton, former director of the United Kingdom Meteorological Office

---

As a journalist, I've had my scientific bona fides called into question by those convinced climate change is a grave, modern concern – as though writers have not studied different subjects and then discussed what they learned in print before. I would like to point out that it is climate scientists who have truly left their own realm and come stomping onto *writers'* turf – that of narrative.

The climate scientists who have delved into this particular narrative the most successfully are, for the most part, native English-speakers. Indeed, global warming is a tale that is principally written in English. Scientists from a number of non-English-speaking

Dominance by English-speakers in the worldwide conversation about climate is, perhaps, an echo of each nation's imperial past. *The Union Jack and the Stars and Stripes.*

countries have added a detail or two along the way, but no one can compete with the influence of British and American scientists in the story's telling. If you are one of the people who fears the effects of manmade global warming, and your native language happens to be English, you might want to think about that for a moment or two. One reason for the disproportionate power of English in the master narrative is the simple fact that the longest continuous record of temperature anywhere on Earth comes from England. Known as the **Central England Temperature**, or CET, the record extends back to 1659. An aggregate of four temperature stations falling within a triangle covering a sizable hunk of England, CET is its own form of scientific achievement.

As for other aspects of the British scripting of the story, the United Kingdom is also home to several institutions and figures that have held enormous sway in the climate debate. These include the **UK Met Office**, the **Hadley Center for Climate Prediction and Research**, the **Climatic Research Unit** at the University of East Anglia, and, counter-intuitively, Margaret Thatcher.

The United Kingdom had been the site of some of the earliest (and quite important) air-pollution legislation, beginning in the 1870s. In the 1950s, the issue came to the fore again with legendary pea soup fogs laced with poisonous gases and particles, leading to the development of a significant environmental movement as well as further legislative curbs on emissions. In 1970, the "greenhouse effect" was mentioned in a debate at the Houses of Parliament for the first time. A year later, the Climatic Research Unit was instituted at the University of East Anglia. Its first director, Hubert Lamb, was of the opinion that *global cooling* was the principal concern in terms of a human effect on the atmosphere. The institute's second director, Tom Wigley, oversaw a change of focus away from global cooling and toward global warming, during a tenure that lasted from 1978 to 1993.

In the late 1970s and early 1980s, Britain was facing a prolonged energy crisis, partly brought on by a series of coal-mining strikes. When the Conservatives won the parliamentary election in 1979, Margaret Thatcher was made prime minister and took her mandate to include ending the coal strike by whatever means necessary. Thatcher went after the coal-mining unions with a series of controversial moves including stockpiling coal, converting coal-fired plants to burn oil, and closing 20 coal mines (leading to 20,000 job losses). She also acted throughout her long tenure as prime minister to develop her country's nuclear power capability.

As the 1980s unfolded, Thatcher strategically funded the climatologists alleging grave risks from increasing carbon dioxide in Earth's atmosphere, in order to further weaken the coalminers' position. Her science advisors, led by Nigel Lawson, were, meanwhile, working to show coal to be a non-viable fuel. The principal beneficiary of the Thatcher government's funding of climate science was the UK Met Office, or **Met O**, as it is now known. The Met O began issuing grim proclamations regarding greenhouse gases,

The particular strength of English-speaking scientists within the United Nations owes in part to the enduring power of the United States and the United Kingdom on the world stage. But a percentage of it comes specifically from the political calculus of the United Kingdom's prime minister during the 1980s. *Margaret Thatcher, Washington, D.C., 1977.*

particularly carbon dioxide, giving Thatcher fuel in the political battle over coal. By the end of the decade, extreme weather events were being attributed to manmade emissions by more and more people and institutions, including the Met O. A ferocious storm that the Met O failed to predict in 1987 became a talking point for those alleging human-authored climate change. "We would have warned you of the storm, if it had been of the old-fashioned variety," the Met O seemed to be saying.

Thatcher's choice, in 1983, to oversee the Met O had been John Houghton. A professor of physics at Oxford, Houghton suddenly found himself the chief of a well-endowed scientific organization with an influential role in British, and indeed world, policy decisions. Along with other renowned climate scientists, he also became the prime minister's climatology tutor. Seven years later, Houghton was made director of a new, even more influential wing of the Met O. The new entity was the Hadley Centre for Climate Prediction and Research, and reflected the anti-carbon sentiments of Houghton's famous pupil. By that point, Houghton's personal influence had extended to the worldwide scientific community, having been named as the lead editor of the **Intergovernmental Panel on Climate Change**'s First Assessment Report. Houghton is said to have arrived with the final report in hand before deliberations even began.

Other British scientists, particularly those at Climate Research Unit, were working diligently to establish the fragility of Earth's climate, despite abundant evidence to the contrary. Ironically, CRU has been the beneficiary from the start of major donations by oil companies, including British Petroleum and Royal Dutch Shell, which bowed without much of a fight to the public relations power of the climate-change-touting researchers. This distinguishes CRU scientists and those in their camp from skeptic scientists, not one of whom has been shown to have received similar funding from "Big Oil."

By the 1990s, the older CRU and newer Hadley Center were working in close collaboration, amplifying each other's considerable prominence. By the middle of the decade, a rising star within CRU was Phil Jones, who, like Houghton, would eventually enjoy international exposure and acclaim. Such attention was a heady concoction for an anonymous scientist like Jones, whose research focus had been flow changes in a single English river.

In his initial, quite humble, role at CRU, Jones was a data compiler, working on aggregated land temperatures from all over the globe. This was in order to come up with CRU's portion of HADCRUT, a collaborative product between the Hadley Center and CRU. The Hadley Center had been tasked with compiling sea surface temperatures and Jones and CRU had been tasked with compiling land temperatures. The two datasets were averaged on a monthly basis, both in the present time frame and going back to the 19th century, in order to produce what has come to be known as the **global mean temperature**. There are long, cogent arguments about why such an "average temperature" has no physical basis. Bjorn Lomborg has suggested, for instance, that the "global temperature average" makes as much mathematical sense as averaging all the phone numbers in the world, and that is about right. Nonetheless, the HADCRUT version of this figure became the single most important piece of scientific research ever created, when it was adopted by the IPCC in 2001.

When journalists publish a graph showing rising global temperatures, there is a better than 50-percent chance that the data on the graph are a representation of HADCRUT. Thus, during the 30-year debate that has characterized atmospheric science since the late 1970s, British climatology, to a significant extent, became the United Nation's climatology, both in terms of data and personnel. As the UK Met Office, the Hadley Center within it, and the CRU at the University of East Anglia achieved wider and wider acclaim, through the UN's imprimatur, budgets were dramatically increased at all three. In 2009, the Met O installed the most powerful supercomputer in the United Kingdom. Dubbed "Deep Black," the machine immediately became one of the largest electricity customers in the country, slurping up 1.2 megawatts of energy, or enough to power 1,000 homes.

Some money that CRU took in, starting in the 1980s, was American. In the wake of the Three Mile Island nuclear disaster, the United States Department of Energy (DOE) expanded its budget to look into the various effects of energy production on the environment. The department, at a stroke, became one of the major funders of climatology, which, as a nascent science, relied on the development of datasets that had yet to be created. The Climatic Research Unit was one of the first beneficiaries of the DOE funding, and has continued to receive substantial funds from the United States ever since.

Another reason for the British dominance of climatology is the history of the British Empire. Hundreds of thousands of temperature records taken by Royal Navy ships as they crisscrossed the world's oceans in the mid-nineteenth century, and since, form the backbone of this authority. No other country had such a thirst for meteorological knowledge coupled with the far-flung hardware to satisfy it. While problematic in its own right, measuring the temperature of the oceans is the most important part of "taking the Earth's temperature." Among the problems with recording the sea surface temperature are changes in methodology. For more than a century, ship crews literally dropped a bucket over the side of the ship and put a thermometer in the water that they had retrieved. Now, increasingly, water is taken in via uptake tubes and measured in a wholly different environment. Nonetheless, direct measurement of sea water remains an area of British supremacy.

So, while the saying "Rule, Britannia" has faded in the sense of 19th century colonialism, it has come to hold true in the realm of science, and, by effect, that of international politics – a kind of back door to world dominance. For his part, CRU's Phil Jones, in the run-up to what would become known as Climategate, expressed anxiety about the perception of Western scientists as the latest in a line of colonists. In an August 2009 e-mail exchange he pointed to sensitivities regarding "data imperialism" as reasons for hoarding his culled data with climate skeptics. The idea that such niceties shielded him from Freedom of Information requests, and the requirements of standard scientific practice, seemed, on the face of it, to be a fantasy.

With the strength of the U.K.'s climate institutions alone, English might well have become the de-facto official language of global warming. However, U.K. researchers have had a powerful partner in their counterparts from the United States. As Margaret Thatcher, John Houghton, and Phil Jones established the British Empire in climatology, much was going on in a similar vein on the American side of the pond. Institutions and individuals with prestige and sway to rival those of the UK were stirring by 1975. Among the more prominent scientists focused on climate in the U.S. were Margaret Mead, Stephen Schneider, James Hansen, and Michael Mann. A culture of "you scratch my back, I scratch yours" quickly developed among the American group and their British counterparts. In the process, the scientific method began to take a beating. Among the institutions delivering blows were the United States Department of Energy, a host of universities (led by Stanford and Columbia), NASA, and the National Oceanic and Atmospheric Administration. Although not a scientist, then-Congressman Al Gore provided platforms to the American researchers announcing incipient climate chaos in the early 1980s, ensuring them media exposure and increased funding. Gore's deep-pocketed foundation continues to provide material assistance to the most prominent global-warming scientists and to produce public-relations materials on their behalf.

One can push the origin of the theory of carbon dioxide heating Earth's atmosphere back to the 19th century, when the Swedish scientist Svante Arrhenius began publishing on the subject. In the United States, measurements taken by Al Gore's science teacher at Harvard, Roger Revelle, starting in the 1950s, contributed strongly to the view that industrial activity could cause an out-of-control ocean-atmosphere system. That perspective was increasingly common among American scientists by the early 1970s, albeit prodded by a different concern: particulate pollution. With temperatures falling and Arctic ice thickening, scientists (including many who would later warn of warming) began to alert their governments. It was in this context that Richard Nixon's State Department created an entity known as the Panel on the Present Interglacial. The United States soon issued a report called "A United States Climate Program," and, by 1975, bills had been introduced in the United States Congress to combat the perceived threat of global cooling.

That same year, Margaret Mead, the American anthropologist, put together a conference to address mankind's damaging influence on the atmosphere. At that time a recent president of the American Association for the Advancement of Science and a loud opponent of continued human population growth, Mead called her conference "The Atmosphere: Endangered and

Endangering." Among those attending were Stephen Schneider. Schneider, already near the top of his field, would eventually go on to educate a small army of climate scientists about the perils of global warming. First, though, he spent nearly a decade warning about a coming Ice Age.

Amazingly, despite his pirouette from Ice Age grandstanding to greenhouse scare stories, Schneider gained massive sway in the worldwide media regarding climate change. Part of his success owed to his frequent, vehement disavowals of having warned the world about the Ice Age, but his 1976 book, *The Genesis Strategy*, extolled the virtues of storing food for a looming mega-freeze. He was interviewed by Leonard Nimoy in a somewhat notorious episode of the pseudo-scientific "In Search of…" in 1978 that focused on a recent spate of cold winters and fears that the Great Frost was nigh. Within three years, though, he was instead worried about global warming and, in 1989, having completed his remarkable transformation, he gave a quote to *Discover* magazine about global warming that dogged him until his death in July 2010:

> On the one hand, as scientists we are ethically bound to the scientific method, in effect promising to tell the truth, the whole truth, and nothing but — which means that we must include all the doubts, the caveats, the ifs, ands, and buts. On the other hand, we are not just scientists but human beings as well. And like most people we'd like to see the world a better

place, which in this context translates into our working to reduce the risk of potentially disastrous climatic change. To do that we need to get some broad-based support, to capture the public's imagination. That, of course, entails getting loads of media coverage. So we have to offer up scary scenarios, make simplified, dramatic statements, and make little mention of any doubts we might have. This "double ethical bind" we frequently find ourselves in cannot be solved by any formula. Each of us has to decide what the right balance is between being effective and being honest.

Schneider kept the quotation on his website and confirmed that he was accurately quoted.

Another scientist at the conference organized by Mead was John Holdren, currently the director of the White House Office of Science and Technology Policy. A devotee of the Armageddon-minded environmentalist Paul Ehrlich, Holdren has never yet budged from the view that humankind stands at the edge of a precipice. During the conference, an end-justifies-the-means, bunker mentality gained currency among Holdren and the other scientists, spurred by a credo promulgated by Mead. She said failing to get on board with doomsday scenarios while scientists waited for the research to be completed was the "modern equivalent of fiddling while Rome burns."

At about the same time, a physicist named James Hansen was a few years into a career at NASA's Goddard

Institute for Space Studies, of which he became director in 1981. Hansen's rise at NASA mirrored the agency's own ascendancy in climatology, and was one reason for it. Focused initially on the CO_2-rich atmosphere of Venus, which had helped turn the second planet from the Sun into a hothouse, Hansen subsequently became interested in rising CO_2 in Earth's atmosphere. The cachet of NASA, coupled with Hansen's considerable skill in dealing with the media, has guaranteed the agency and the man significant roles in authoring the global warming narrative.

Along the way, Hansen has made some startling claims. In 1988, he was interviewed in his NASA office within Columbia University. Asked how the scene outside the window would change in 20 years, he said that the West Side Highway would be under water from rising sea levels, the windows in the building opposite would be taped up due to storm damage, and there would be more police cars because of the increased heat-wave related crime. The fact that not a single one of the predictions has been remotely borne out has done nothing to diminish his role in spinning the narrative of global warming, both in peer-reviewed articles and in the mainstream media.

His latest popular book, *Storms of my Grandchildren*, focuses on the destructive strength of global-warming-enhanced storms that he fears will make the world a more dangerous place in the decades to come. But human memory is notoriously short, perhaps nowhere more so than in the realm of storm history. A 1287 storm known in the Netherlands as "St. Lucia's flood," as well as the Grote Mandrenke of 1362, the Great Storm of 1703, and the North Sea Flood of 1953, each occurred before global warming allegedly began, and each was extremely deadly. The 1287 storm killed tens of thousands, and redrew the Dutch and British coastlines. If a storm of such proportions were to occur today, you can rest assured that Hansen would be in front of a microphone attesting to CO_2's destructive power.

Still, no one on Earth (neither in the IPCC nor in any rival institution promoting catastrophe) can compete with Al Gore. Through his books delineating the end-of-the-world through carbon-dioxide-induced warming, through the three hundred million dollars spent by his foundation on promoting the same story since 2008, through the production of one of those books into an Oscar-winning movie in 2006, and through his Nobel-acceptance speech and his statements since, Gore commands the world stage.

As the story of global warming continues to be told by Gore, and by the American and British scientists with whom he is allied, one question remains: How much warming has there actually been? In order for decisions about national economies the world over to be made under "crisis" conditions, must events have already taken place to justify such actions? Not exactly. Even if you grant the more extreme climate scientists the maximum warmth they claim, you're really only talking about 1.5 degrees Fahrenheit of additional warmth over the course of the last century and a half.

Ask yourself this, if the TV weatherman in your area predicted a high tomorrow of 74 degrees and instead it turned out to be 75.5 degrees, would you notice?

The upper end of temperatures shown on this graph have been exceeded dozens of times in Earth's past. *A detail from Michael Mann's "hockey stick."*

7 How Hot Is It?

People don't notice whether it's summer or winter when they're happy.

– Anton Chekhov

You would want to believe that if scientists were going to dare to embroider a horror story, in the horror-rich language of English, one about rapidly escalating temperatures and vicious new weather as a result, that those scientists would know, at a minimum, what the temperature of Earth's atmosphere was. However fervently you wanted to believe this, however, you would be wrong. Measuring the temperature of Earth's atmosphere in both the present and the past is enormously complex. Recognizing this led me to a simple question: What is the global mean temperature?

This is a very difficult question to answer, it turns out. For one thing, measuring the temperature in or near a city is not a simple matter. Extrapolating from the temperature of that single city to that of all the cities in even a single country among the hundreds of countries that circle the globe is a daunting task, one to which science may not yet have risen. And while many Americans, having heard about the horrors of global warming, will tell you that they can *feel* the increased heat themselves, it is perhaps worth noting that the two warmest years on record for the United States, in a near statistical tie, are 2006 and 1934. So, if you thought in the middle of the decade just ended that temperatures were getting warmer in the U.S., you were right – they had become as warm as they were 72 years before.

For some reason that simple fact was not mentioned in Gore's film. Ironically, for all the technology visible in the movie, and for all the United States' technological strength and scientific knowledge, the American temperature-recording network is, for lack of a better word, shoddy. This is bad news, because it is probably the best in the world. The reasons the U.S. system is probably the best in the world are multiple: geographic breadth, station maintenance, and frequency of upgrades.

Nonetheless, in 2006, a man by the name of Anthony Watts, a meteorologist with a local radio station in Northern California, saw fit, out of simple curiosity, to go and look at the way his hometown weather station recorded temperature. What he found surprised him.

To understand why he was surprised, it helps to know a little bit about how temperature is supposed to be measured. The most widely used device approved by the National Weather Service for use in the United States Historical Climatology Network is a Stevenson screen. This is a white wooden box mounted on four narrow legs. It has slatted sides and a slatted door to prevent the interior from becoming an oven. There are rules governing the proper siting of a Stevenson screen. It is to be mounted about five feet off the ground. It is to be located more than 100 feet away from buildings. In the Northern Hemisphere, the face of the box is to be pointed away from the south, to minimize sun exposure of the thermometer when opening and closing the door to take readings. It is to be on a natural surface, preferably grass. It is not to be mounted over asphalt. It is not to be mounted anywhere near heat sources of any kind. Finally, it is required to be painted with white latex paint.

The station that Watts examined was on the California State University farm. It consisted of a Stevenson screen, correctly painted in white latex, and appropriately sited. When Watts opened the door, however, he found that along with an electronic thermometer there was a radio transmitter, six inches away. This transmitter, installed to send temperature readings to another location, generated heat. Watts was surprised, but he chalked it up to someone not knowing what they were doing and presumed it to be an oddity.

The second temperature station that he looked at was in the town of Orland, about twenty miles west of Chico. The temperature station in Orland checked out better. It had traditional maximum and minimum mercury thermometers, was at the required height over the ground, had been properly painted, and was more than 100 feet away from the nearest building. What he found reassured Watts, until, that is, he looked at a third station about an hour's drive south.

The Marysville station was of another kind, entirely,

something known as a maximum-minimum temperature system, or MMTS, which looks like an upside-down stack of white dinner plates sitting atop a pole with a slightly larger white disk on top. Although apparently more modern than Stevenson screens, MMTS devices have their own set of issues. And, indeed, some technicians who work for the National Weather Service have created their own anagram for MMTS: "Mickey Mouse Temperature System."

While photographing the Marysville MMTS, Watts noticed hot air blowing on his back. Turning around to take a look at its source, he noticed a few things. It turned out that the town had leased space near the temperature station for use as a cell-phone tower. Near the tower, the telecom company had also constructed a small building to house electronics, which, in due course, required air-conditioning. Significantly, the air-conditioning exhaust was directed at the MMTS device. Equally worrisome, the device was in the middle of the town's emergency services parking lot. This created a pair of issues. First, there was heat from the asphalt itself. Second, town employees left their vehicles, with their heat-emitting engine blocks and radiators, parked around the sensor. Any one of these heat sources was enough to raise the air temperature around the device, as Watts duly noted in a blog that he had begun. "It is the opinion of the site surveyor," he wrote, "that the data produced by this station is biased in so many ways that it is essentially useless."

He began taking long drives to look at other stations

Installed in an incorrect location, at an incorrect elevation, with multiple heat sources contaminating the air sample, this is one among hundreds of incorrectly sited stations in the U.S. *The United States Historical Climatology Network weather station in Roseburg, Oregon.*

within California – and, eventually, out of state, jotting down what he found on his blog. At considerable personal expense, he logged hundreds of hours on the road trying to learn more about how the U.S. was recording temperature. Eventually, he enlisted dozens of volunteers, who appeared by the dozen, having read on the blog about some of the odd things that he was finding. Before long, Watts and the volunteers were finding even more startling instances of poor siting. There were Stevenson screens with broken sides, screens that opened toward the Sun, and screens mounted too close to the ground. There were temperature-recording stations in parking lots, at sewage-treatment plants, at airports – all with heat-contamination issues, frequently more than one.

Airports are a special case, so much so that it has been argued that they have a detectable effect on the global climatological record. The issues are multiple. With the rise of aviation, what started several decades ago as little more than dirt landing strips have become great seas of concrete, with airplane, vehicle, and air-conditioning exhaust, as well as heat emitted by ever-expanding and more numerous buildings wafting over the temperature sensors. Graphs of such records generally show a worrisome increase in warmth, unless you know what happened around the thermometers. On the other hand, as is the case where I live, some airport thermometers remain relatively accurate measures of the rural environment.

Seventy of the temperature stations that Watts and his team evaluated, out of a network of twelve hundred, were found at sewage-treatment facilities. As infrared photography confirmed, abundant heat was being discharged by the treatment process, and the artificially warmed air was enveloping the temperature sensors.

Several other stations had steel trash-burning barrels within a few feet of them. Others were crowded by barbecues. Dozens of sites were closer to buildings than the United States Historical Climate Network code permitted. In the three years of his inventory, Watts personally assessed more than a hundred stations. As of late 2010, Watts and his team had visited 948 of the 1,221 stations in the network.

The results? Three percent of all the stations had no red flags associated with them. These received a designation of "Best." Eight percent had only one or two red flags, were found to provide high-quality data, and were judged "Good." Twenty percent of stations had multiple issues, or a single issue that compromised their readings. These were judged as "Fair." Fifty-eight percent of stations were judged as "Poor," having major issues and providing patently low-quality data. Finally, eleven percent of stations fell under the category of "Worst," with unacceptably corrupted temperature readings. Following NOAA criteria, these stations were assessed as having error values beyond ten degrees Fahrenheit.

Just about all of this distortion stems from what are called heat islands. These are areas of locally warmed air that can extend in size from a few feet to many miles across. Larger such concentrations of heat generally fall within densely populated areas and are called **urban heat islands**. Because it lies within an urban heat island, the temperature sensor in Central Park, for instance, records values as much as 15 degrees Fahrenheit higher than the thermometers measure just 10 miles away. Manhattan's tall buildings are, essentially, giant, outdoor radiators. Additionally, the borough's expanses of asphalt, air-conditioning exhaust vents, and vehicle exhaust collectively warm the Central Park thermometer dramatically. The heat is real, and the temperature is real, but neither of them is a reflection of climate, or even weather.

Somewhat surprisingly, populations of a few hundred people have been shown to raise temperatures by several degrees Fahrenheit as well. Correcting the global temperature record to deal with such heat islands, large and small, is an immensely complicated process. And it is one that James Hansen and NASA have been reluctant to discuss. It is thought that the correction applied to GISTEMP by Hansen and his team is in the vicinity of a quarter of a degree Fahrenheit. Watts and others maintain that this is grossly insufficient.

The perception that the urban-heat-island effect complicates the global warming narrative in ways that some scientists do not like is hard to shake. Even where I live, central Austin's heat island affects the perception of weather and climate on a continual basis. On cold winter mornings, for instance, the central Austin temperature is frequently twelve degrees Fahrenheit warmer than the reading taken at the airport. Indeed, at the airport in the autumn of 2010, record low readings occurred on more than eight occasions, but nary a soul heard the news.

Urban heat islands have an unknown influence on the HADCRUT temperature product. Phil Jones, too, has declined to address the issue in interviews, despite repeated requests for comment. Having receiving tens of millions of dollars of U.S. taxpayer money and slightly lesser sums from his own government, Jones emphatically declares that no one should expect him to share data or the computer code for his climate models and number-crunching for HADCRUT, including the urban heat island adjustment. In an e-mail exchange in August 2009, he wrote: "I work in

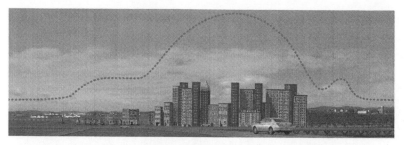

Temperatures in urban areas have climbed due to a proliferation of heat sources. The extent to which localized heat has contaminated the global temperature record is unknown. *Artist's rendering of the urban heat island effect.*

a University. In the UK I am not considered a public servant." In published papers, Jones has strenuously maintained that the urban heat island effect on global mean temperature records is negligible.

Meanwhile, heat islands can and do extend far beyond the bounds of a single metropolitan area. Irrigation is among the ways that giant heat islands are created in such places as California's Central Valley, with such altered temperature environments falling under the category of changes in what is called **land use.** Land that was formerly pale desert (such as in the aforementioned valley as well as in and around Las Vegas, Nevada, and Phoenix, Arizona) is planted and irrigated, suddenly changing the amount of sunlight absorbed. Equally significant, the humidity rises enormously, causing nighttime temperatures to become elevated. Anytime you hear of a "maximum low" temperature, pay attention to where it was measured. In a majority of cases, the warmth is manmade, localized, and unrelated to greenhouse-gas "global" warming. In other words, there are hundreds of locations whose data needs either to be discarded or heavily weighted.

Ironically, Hansen's team at NASA sits within one of the most significant urban heat islands in the world in New York City. I admit that occasionally I wonder whether Dr. Hansen's perception of an increasingly warm world unconsciously mimics his own movement from relatively frigid Iowa to relatively balmy Manhattan.

Being the most technologically advanced nation on Earth, the United States has at its disposal a series of satellites that NASA put into space *specifically to record temperature*. James Hansen points to this capacity on his personal website with pride:

> One of my research interests is radiative transfer in planetary atmospheres, especially interpreting remote sounding of the Earth's atmosphere and surface from satellites. Such data, appropriately analyzed, may provide one of our most effective ways to monitor and study global change on the Earth.

Despite this impressive statement, Hansen and his staff make virtually no use of the data provided by satellites. Like HADCRUT, the NASA global mean temperature product is derived almost exclusively from aggregated individual recording stations on land and at sea. Two other teams – Remote Sensing Systems in California, and the University of Alabama-Huntsville's Earth System Science Center – do derive temperature datasets from satellite readings. NASA, somewhat incredibly, does not. It's enough to want to make one read and re-read Hansen's statement. Although Hansen (a) is embroiled in the climate debates waging around the world, (b) has a singularly high profile among American scientists, (c) claims on his personal website that he uses satellites extensively, and (d) works for NASA, the global mean temperature produced by him and his team mostly comes from the appalling temperature stations such as the ones Anthony Watts has examined firsthand.

If NASA only relied on ground temperature stations from within the United States, but used satellite data for the rest of the world, that would be one thing. In fact, NASA uses an even more patched-together worldwide network of temperature stations, known as the Global Historical Climatology Network to complement its domestic network. One strike against the global network is that a great many countries are represented in it by only a single temperature recording station. That station, almost inevitably, is at the country's largest airport. As Austin Bergstrom International Airport's temperature record exemplifies, some airports' temperature records remain largely uncontaminated by urbanization and airport growth. These are arguably the exception that proves the rule, however.

Meanwhile, some whip-smart and well-informed scientists believe that the very notion of a global mean temperature is a fantasy. The task requires, at a minimum, performing myriad complex statistical tasks. Climatology, by its nature, is a matter of

While mainstream climate scientists insist that skeptics have far more skill in dealing with the media, many have been interviewed by dozens of journalists and had their predictions laid out in their entirety hundreds of times, with scarcely a word of editorial questioning. *Michael Mann, left, of The Pennsylvania State University, creator of the "hockey stick graph," and James Hansen, right, of the Goddard Institute for Space Studies.*

statistical gymnastics, no matter which side of the ideological and scientific fence one may be standing on. That is among the reasons why the arguments taking place about climate in scientific circles, and they are red-hot, are unlikely to be resolved anytime soon.

Michael Mann's graph of historic temperatures, widely known as "the hockey stick," is the most famous case in point. Used extensively by the IPCC, the graph of Northern Hemisphere temperatures shows modern values higher than those of the last 1,000 years, with the older values forming the long portion of the stick, and the modern up-tick forming the blade. Why does the blade of the stick go up so far, so fast? One reason, among many, is that a goodly portion of the data on the graph come from ground-based stations. A second is that heat islands are seldom static. Generally, where there is one, it is growing in size and intensity, as are the majority of the world's airports and the majority of the world's towns and cities. So, all things being equal, if one were to take temperature readings from a system that drew upon *only* stations in urban settings over the past century, one would expect to see a slow and steady but significant warming signal. Although rural stations do constitute part of the Global Historical Climatological Network, the number of non-urban stations used has declined during the past 20 years. A number of the rural stations no longer being used had been in the former Soviet Union, with a preponderance of them in Siberia. When the Soviet Union fell apart, the administrative system that made collating station data possible fell apart as well.

Surprisingly, the most consistent and intense warming signal on Earth, according to NASA, comes from Siberia. Another reason for the relative warming in the former Soviet Union is the fact that observers were known under Soviet rule to under-report temperatures. The low-balled numbers were sent to Moscow because energy allocations were made based on past climate records, with observers attempting to get a little more heat for their own cities and towns.

Has James Hansen gone to Siberia to look at any of the stations showing significant, ongoing regional warming? The answer is no. As Schmidt, his spokesman, has said, the thought that NASA scientists have the

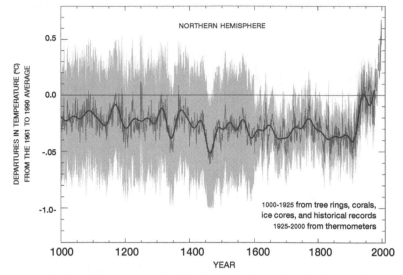

THE HOCKEY STICK TEMPERATURE GRAPH

NORTHERN HEMISPHERE

DEPARTURES IN TEMPERATURE (°C) FROM THE 1961 TO 1990 AVERAGE

0.5

0.0

-.05

-1.0-

1000-1925 from tree rings, corals, ice cores, and historical records 1925-2000 from thermometers

1000   1200   1400   1600   1800   2000

YEAR

Anomaly temperature graphs like the one above, in which one degree Celsius is made to look like a horrifying spike in measurements, are inherently shocking. *Michael Mann's "hockey stick," showing temperature in the Northern Hemisphere since the year 1000.*

time to perform such firsthand research is "laughable." Meanwhile, a high percentage of rural stations either do not follow the upward trend shown by NASA globally or do so at a much-diminished level. Dozens of rural stations' raw datasets show *declines* during the past century.

Measuring temperature in urban heat islands *consistently* is a complex process. Published papers indicate that even a village with 10 people living in it can produce an urban heat island of a degree and a half Fahrenheit, that a town with a population of 100 can create an urban heat island three degrees Fahrenheit warmer than the surrounding countryside, that for a town of 1,000 inhabitants the warm bias is more than four degrees, and that for a city of a million people the

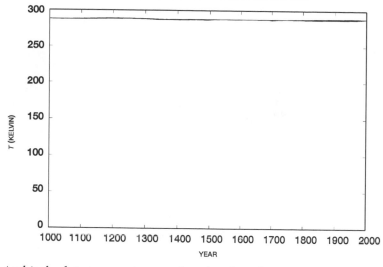

EARTH'S TEMPERATURE EXPRESSED IN KELVINS

As this absolute-temperature graph makes clear, the temperature of Earth's atmosphere since the year 1000 has hovered relatively near 287 Kelvins, which is about 14 degrees Celsius, and about 57 degrees Fahrenheit.

• •

warm bias is just shy of 10 degrees. This is where things get really convoluted. Hansen routinely changes *historical data*, figures written down by observers as much as several decades in the past. Downward adjustments of historic readings and upward adjustments of modern readings produce visibly steeper upward trends on temperature graphs.

The case of a single station – in Lampasas, Texas – is instructive. First, a station move from a reasonably reliable location over grass to a position in a parking lot, next to a six-foot black satellite antenna, and eight feet from a two-story brick and plaster building with multiple air-conditioning units in the windows. Taking a glance at the Lampasas record, one can easily discern the never-before-attained values in the early 2000s following the move.

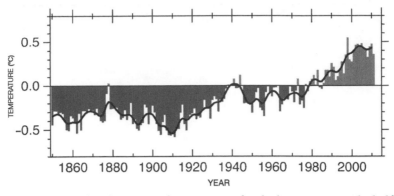

GLOBAL TEMPERATURE SINCE 1850

According to this depiction of temperature for the last century and a half, both the rate of warming and the amount of warming from 1910 to 1945 are similar to what occurred from 1975 to 1998. *HADCRUT temperature from 1850 to 2011.*

- -

But the station move is just part of the problem with Lampasas. Another is the historical adjustment made by NASA. On top of the skyrocketing data resulting from the station change, Hansen lowered the past historical data by about a half-degree Celsius, in the name of something called homogenization. The effect was to take an artificially upward trend, one of the most dramatic in the temperature network, and make it more so.

Similar adjustments abound, Anthony Watts has found. "There is a quality-control problem all the way through the system," he says. "One of the stations that we located is what is called a lights-equal-zero station in Cedarville, California." The designation, Watts explains, means that the station has not seen any urbanization whatsoever, no population growth, no streetlights installed, no heat island. That being the case, one would assume that if NASA changed Cedarville's data within an increasingly urbanized

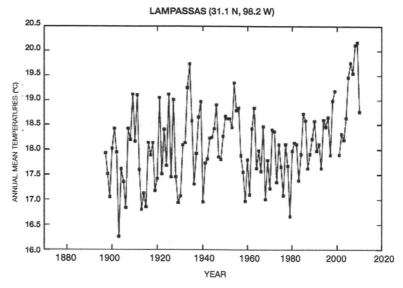

LAMPASSAS (31.1 N, 98.2 W)

Every temperature station has a story. Without knowing about a change in location of this station's thermometer casual observers might falsely conclude that global warming was visible in its record. *Goddard Institute for Space Studies temperature record for Lampasas, Texas.*

network the adjustments made would be toward the cool end of the spectrum. In fact, Watts says, "The temperature was actually adjusted upward. Why would a station that is out in the middle of nowhere have its temperature adjusted upward?"

Unfortunately, Cedarville's NASA-altered data are not unique. More than a few stations with obvious heat-contamination issues have had their values *raised* by Hansen's team. The Canadian climate blogger Steve McIntyre has called Hansen on this with regard to a station in Detroit Lakes, Minnesota, and has been repeatedly insulted by NASA staffers as a result. After months of denials, it turned out that there was a data splicing error that NASA had made. McIntyre, for the record, never received an apology.

Partly because of the prestige of NASA itself, partly because of James Hansen's personal relationship with Al Gore, partly because the United States is *presumed* to be the most technologically advanced nation on Earth, Hansen's data, although a statistical outlier among the four top global mean temperature providers, is influential with regard to the United Nations and its Intergovernmental Panel on Climate Change. He is correct to observe a trend, of course. The world has warmed, somewhat, during the last 30 years, but it has been warming, at pretty much the same rate, for the last 200 years, since the end of the Little Ice Age.

As for the sudden, frightening warming on Gore's graph in *An Inconvenient Truth*, at least some of it originated in urban heat islands and other temperature-recording problems. Although Watts and Hansen have never met, they probably should. It is fair to say that Watts knows *a lot* more about the temperature station network that Hansen uses to get his numbers for United States temperatures than Hansen does himself.

How can this be? Partly, it is a function of bureaucracy, the layers of which are many when it comes to the United States' temperature station network. Directly supervised, to the extent that they are, by the National Weather Service, the stations are largely volunteer-operated, part of a system that came into being at the turn of the last century. Some stations have been automated, such as the one in Chico that Watts first examined. Many others have not. Manual

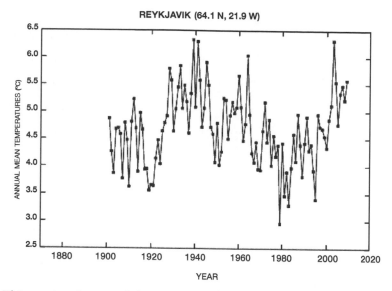

REYKJAVIK (64.1 N, 21.9 W)

This station is one of dozens around the globe to show either no warming, or cooling, during the last half-century. Warming during the 1920s and 1930s is easily seen. *Goddard Institute for Space Studies temperature record for Reykjavik, Iceland.*

temperature recording alone introduces an element of uncertainty into the data stream, although for the most part the volunteers are admired for doing a hard job well.

Once observers take down their information, it is relayed to the National Climatic Data Center (NCDC) in Asheville, North Carolina, an arm of the National Oceanic and Atmospheric Administration. NCDC looks for irregularities and otherwise grooms the data before turning it over to NASA. Any time Hansen's team at NASA's Goddard Institute for Space Studies publishes numbers representing the global mean temperature, questionable measurements have been turned into a mathematical sausage of ghastly origin. As Hansen's spokesman Gavin Schmidt has written, regarding the ability of NASA scientists to measure

1986	1.51
1987	2.33
1988	2.09
1989	1.27
1990	1.31
1991	1.02
1992	0.43
1993	1.35
1994	1.90
1995	1.98
1996	1.19
1997	1.96
1998	2.93
1999	0.94
2000	1.74
2001	1.59
2002	2.56
2003	2.27
2004	1.57
2005	2.53
2006	1.72
2007	2.14
2008	1.66
2009	1.88
2010	2.42

Amount of increased carbon dioxide in the global atmosphere, in parts per million. Notable years: **1992**, which saw the smallest increase in the record, due to cooling produced by the eruption of Mt. Pinatubo, and **1998**, which saw the largest increase in the record, due to warming produced by the Super El Nino.

temperature themselves and make firsthand sense of readings, "The idea that they are in any position to personally monitor the health of the observing network is laughable."

So, rather than ask the deceptively simple question "How hot is it?" one would be right to ask instead, "How hot is it at rural stations?" Dozens and dozens of such temperature records from the United States and countries around the world show effectively no rise during the last century. As for the satellite-measured warming of the global mean temperature from 1978 to 1998, which has not continued during the last thirteen years, there are a variety of factors that could have caused it.

The face of the Sun during periods of low activity has few sunspots, or none. If the predicted period of low solar activity continues to unfold, Earth's ocean-atmosphere system may cool.

8 The Quiet Sun

*I expect the tempest over the mountains of the moon will
be a joke compared to the lashings I will receive over
these clouds.*

– Galileo, referring to his study of sunspots

Just as climate scientists have controlled the Master
Narrative in such a way as to raise public awareness of
the power of carbon dioxide to affect the temperature
of Earth's atmosphere, and to exaggerate the warming
of the past few decades, they have typically fought to
tamp down awareness of another feature of the
climate system. That feature? The Sun. You can start a
conversation with anyone you might meet at a bus
station about the power of CO_2 to seriously erode the
environment and amplify everything bad about the
weather. But try having a similar conversation about
the Sun. You're as likely as not to receive a blank stare
for your efforts.

Not a single one of the dozens of computer models
forecast the nearly flat global mean temperature from
1998 to present, despite the hundreds of millions of
dollars poured into the project. Part of the reason for
the computers' miscue was that they had not been

programmed with sufficient emphasis on our Sun. On the other hand, with an overly linear conception of the role of carbon dioxide in raising atmospheric temperature, how could the machines do better than they have done? Another reason that the models' predictions of Earth's climate are so inaccurate is that the individual sciences upon which they depend cannot generate accurate predictions within their own realms. Nonetheless, Al Gore bases his conclusion almost entirely on such computer-model fantasies and insists, "The science is in." At best, this is ignorance.

Among all the areas of climate study in which the science is not in, likely the most significant is **solar variability**. Generally, the Sun follows an 11-year cycle between high and low levels of activity. The period of high activity is characterized by large numbers of sunspots darkening portions of the Sun, a relatively significant number of high-speed particles being emitted in the **solar wind**, and frequent **solar flares** – sudden eruptions of magnetic activity that can, in their most extreme form, knock out communications on Earth. The period of low activity is characterized by the opposite set of conditions: the face of the Sun often has few or no spots upon it, the solar wind quiets down, and solar flares become relatively rare. One cycle of solar activity can be as brief as eight years and as long as 14 years. The longest cycles have extended quiet periods at their conclusion, and some solar physicists argue that these long quiet periods allow both primary and secondary cooling effects to kick in.

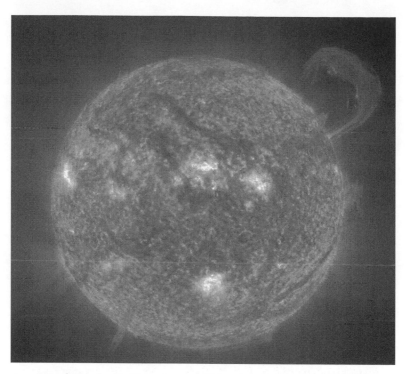

A coronal mass ejection, upper right, in an image captured by a satellite observatory, suggests the dynamic nature of our nearby star.

. . . .

Whether long or short, **solar minimum** is when the face of the Sun is blank of sunspots, or nearly so. Conversely, during **solar maximum,** dozens of black spots can dot the Sun's face. Sunspots, which typically form in groups, are cooler-than-normal regions on the surface of the Sun. Each sunspot is a magnetic storm. Out of these cool, magnetically active freckles come the bursts of energy known as solar flares. Flares, like sunspots, are strongly magnetic, and affect Earth's three-dimensional **magnetic field**, known as the **magnetosphere**, pushing it down on the Sun-side and stretching it out on the shadow-side.

Solar minimum can also refer to a *series* of weak

cycles, such as during the **Maunder Minimum** – a famous period of no sunspots from 1645 to 1715. Likewise, solar maximum can refer to a series of strong cycles, such as the **Modern Maximum** – a period of unusually high solar activity from the late 1940s until the mid-2000s.

Contemporary solar physicists around the globe attempt to foretell the strength of coming solar cycles. In the process, they are joining a long line of scientists who have closely studied the Sun, many of whom have been belittled and ostracized because of their interest. For the idea of solar variability has been considered heretical at least since Galileo and others developed the telescope in the early 17th century. As Galileo's biographer James Reston and others have written, the Grand Inquisitor was far from happy about the observations of sunspots that Galileo wished to publish. The Inquisitor proposed that the astronomer *might* want to consider the possibility that the inky splotches were instead "pure stars moving across the sun." The principal problem for the Church was that Galileo's observations indicated a rotating Sun, as opposed to the still and perfect orb posited by papal scholars. In the end, Galileo saw that the spots were not clouds at all but marks on the Sun's face, and he was unwilling to take the Inquisitor's hint. While he does serve as a scientific role model for both the quality of his observations and his bravery, he did pay a high price – spending the rest of his life under house arrest.

The idea that sunspot-related solar variability

influences Earth's environment is frequently attributed to an early nineteenth-century British astronomer by the name of William Herschel. His observation was that a relatively high number of sunspots correlated to larger wheat crops, and lower prices. Although his study was widely mocked, peer-reviewed research in the last decade has partly borne out his hunches. The European Space Agency named its new satellite observatory after Herschel in May 2009.

At the turn of the 20th century, another English astronomer, Edward Maunder, conducted a ground-breaking study of solar cycles. He was especially interested in periods of low sunspot activity, such as those that took place from 1420 to 1550 and from 1645 to 1715. Although completely ignored by his peers, Maunder's work was eventually built upon in the 1970s by an American solar physicist by the name of Jack Eddy. Eddy studied solar cycles extensively himself, and matched carbon-dated temperature proxies with them, to good effect. It was Eddy who labeled the coldest period during the Little Ice Age the Maunder Minimum, and his work has been influential. NASA graphs of past sunspot cycles that include the Little Ice Age typically now show the Maunder Minimum.

NASA has been taking measurements of solar activity since the early 1970s, initially in preparation for a mission known as Skylab. The two leading reasons for the agency's interest in solar science are: (a) the amount of radiation that astronauts are exposed to, which rises and falls with the solar cycle and (b) those who run telecommunications networks and power grids require advance warning of solar storming, which can bring down both kinds of systems. In the three decades since Skylab was put into space, curiosity about the Sun's effect on climate here on Earth has evolved for many working at the agency. In 1990, James Hansen wrote that "comparisons of available data show that solar variability will not counteract greenhouse warming." Conversely, in 2009, he wrote, "it is likely that the sun is an important factor in climate variability."

As Hansen moved toward this shift in perspective, a Russian space program scientist had come out swinging, boldly predicting that Hansen's forecasted warming during the next half-century would simply not come to pass. In a 2007 interview with the Russian news agency RIA Novosti, Habibullo Abdussamatov, the head of the solar science wing of the Russian space program, said that a grand solar maximum had recently ended and the consequences would be apparent soon. "Instead of professed global warming, the Earth will be facing a slow decrease in temperatures in 2012-2015," he said. "The gradually falling amounts of solar energy, expected to reach their bottom level by 2040, will inevitably lead to a deep freeze around 2055-2060." He stated that the recent uptick in global mean temperature "results not from the emission of greenhouse gases into the atmosphere, but from an unusually high level of solar radiation and a lengthy – almost throughout the last century – growth in its intensity."

Abdussamatov was joined by several solar physicists

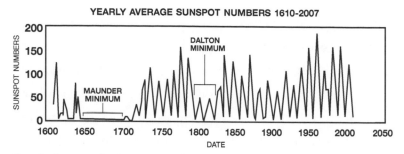

YEARLY AVERAGE SUNSPOT NUMBERS 1610-2007

Graph showing fluctuations in the number of sunspots on the face of the Sun. Several features are worth mention: (1) The Maunder Minimum from 1645 to 1700, during which sunspots ceased to be visible on the Sun and temperatures on Earth were the lowest they had been since the last full-fledged ice age; (2) the Dalton Minimum from about 1790 until about 1830, during which temperatures on Earth fell again; and the high number of sunspots during the grand solar maximum of the middle to late twentieth century, when temperatures on Earth rose to their highest levels since the Medieval Warm Period.

· ·

noting the high level of solar activity during the last several decades. Sami Solanki published a paper in 2004 entitled "Unusual activity of the Sun during recent decades compared to the previous 11,000 years." In it, he argues that the Modern Maximum was the greatest period of solar activity during the Holocene, by a wide margin. For his part, Abdussamatov went on record that he thought conventional notions about global warming were mistaken: "[T]he common view that man's industrial activity is a deciding factor in global warming has emerged from a misinterpretation of cause and effect relations."

Over the past few decades, solar physics has drawn more researchers into its controversy-laden midst. In the meantime, the peaks and valleys in historical solar activity, made decipherable through a combination of direct and indirect measurement, have all received

159

names. For the quiet periods, they are: the Oort, which took place between 1040 and 1080; the Wolf, which took place between 1280 and 1350; the Spörer, which took place from 1420 to 1550; the Maunder, which took place from 1645 to 1715; and the Dalton, which took place from 1790 to 1820. The named periods of high activity are: the Medieval Maximum, from 1100 to 1250; and the Modern Maximum, which began in 1950 and ended in 2006. The Sun-Earth climate connection scientists avow that the named solar minima, starting with the Wolf and ending with the Dalton, were almost surely among the causes of the Little Ice Age. What Abdussamatov was claiming was that a repeat of the Maunder Minimum was upon us, with similar kinds of cooling in store.

Another factor that may have altered Hansen's views was the development of theories about secondary solar effects, especially one by Henrik Svensmark. In a campaign to find the holy grail of climatology, something that could explain many of the ups and downs during the Holocene as well as over the course of deep geologic time, Svensmark had come up with a mechanism that included **cosmic rays**. Scientists who said that the Sun's main output of energy was far too constant to explain significant climate variation on Earth were failing to take into account knock-on effects, specifically that cloud production was largely regulated by the Sun's influence on cosmic rays, Svensmark and those in his camp have maintained that more clouds should produce cooler temperatures.

Probably more significant in terms of convincing the doubters to at least pay lip service to the importance of solar variability, the Sun had begun to do what Abdussamatov had said it would, by entering a prolonged minimum, the deepest in at least a century.

Somewhat ironically, as the downturn in solar activity began, physicists at NASA (not the wing overseen by Hansen) were forecasting a period of intense solar activity. In March 2006, as the solar cycle began winding down, NASA, on behalf of its solar physicists, put out a somewhat breathless press release claiming that the coming cycle was going to be "a doozy." The release focused on a National Center for Atmospheric Research scientist named Mausumi Dikpati who said that the high level of solar activity during the last half of the twentieth century was about to be built upon. "The next sunspot cycle will be 30 to 50 percent stronger than the previous one," Dikpati said. Basing her prediction on something called the "solar dynamo" theory, in which a "conveyor belt" moving magnetic fields from the poles toward the Sun's equator drives sunspot formation, Dikpati and her team also relied heavily on computer modeling.

NASA's own leading solar physicist David Hathaway, concurred with Dikpati that the coming cycle would be significant, predicting a number of sunspots among the highest ever seen. The only exception Hathaway took with Dikpati was on the timing. "History shows that big sunspot cycles 'ramp up' faster than small ones," he said.

The next two years were anything but kind to

If the downturn in the Sun's activity is prolonged, the grand solar minimum may be named after the man whose research brought the Maunder Minimum to the world's attention in the 1970s. *The late John A. "Jack" Eddy, in a photo made public by his widow, Barbara Eddy.*

Hathaway and Dikpati, however, as the old cycle showed no sign of ending and the new cycle no sign of starting. In what some labeled punting, Hathaway reissued predictions every several months, but kept with his overall view that a strong cycle was on the way.

As U.S. and Russian space-program physicists took part in a tacit competition to anticipate what the Sun would do in the coming decades, the Space Race of the 1950s and '60s was arguably being played out anew. Unlike the first time, the competition was being held out of public view, with the countries' best physicists, Hathaway, Hansen, and Abdussamatov among them, trading blows in academic journals. This time around, the terms of the debate were not rocket-booster fuel formulae, re-entry material construction, and orbital equations, but, rather, where the planet was headed climatologically.

Jack Eddy, at the end of his life, was paying attention to the Sun's doings. "We're at a prolonged minimum now, of which there have been precedents," he said in 2008, months before his passing. "Whether we're going to go into one of these profound minima or not, we won't know until we get there. It might make me famous if it happens, but I don't see that we know that it will happen." Eddy's work has been cited by Abdussamatov, and indeed calls have been issued for the naming of the forthcoming minimum, should it be as prolonged as some anticipate, as the Eddy Minimum. It is a matter of some irony that a Russian scientist following his research to its logical conclusion appears to have greater intellectual freedom than most of his American counterparts, with the notable exception of Jack Eddy.

Assuming that Abdussamatov and the Russians are correct, the geopolitical consequences will likely rival those that sprang from their country's putting the first astronaut into space a little less than 50 years ago. NASA's subsequent explosive growth, and partial transition into a climatology center, owe, to some extent, to the lifting of cosmonaut Yuri Gagarin into space on April 12, 1961. On that occasion, the United States' response, under the leadership of President John F. Kennedy, was swift. In a speech to a joint session of Congress, JFK announced America's intention to visit the Moon, a plan that bespoke (a) high ambition and (b) deep U.S. embarrassment about having been largely outclassed, up to that point, by Russian science and technology.

In the current case, competing views on climate have already had cascading effects on world energy, carbon trading, and geopolitics. In 2004, Vladimir Putin surprised many by reversing course and ratifying the Kyoto Climate Protocol. His reasons for doing so were canny. First, Kyoto was linked to 1990 emissions levels, and Russia in 1990 was in the midst of economic disaster following the breaking up of the Soviet Union. What that meant in practical terms was that not only was it easy in 2004, and for the foreseeable future, for Russia to meet emissions-reduction targets, but that the country would have emissions credits *to sell*, at hefty annual profit, to western nations whose economies had grown substantially in the intervening years.

Until deciding that Kyoto was good for his country, Putin dismissed global warming out of hand. He was backed in this by a cadre of Russian scientists who rejected flat-out American and European theories about accelerating global warming. Simultaneously, Putin was consolidating state control of briefly privatized Russian energy firms, chief among them Gazprom. Putin knew that if his own science advisors were correct, the Europeans' need for Russian natural gas and petroleum would not diminish anytime soon – just the opposite. With this gambit, Putin could increase his political and financial muscle, and make the Motherland a little side money on carbon trading.

As Putin placed his hand on the spigot of Russian petroleum and natural gas, he began to let former Soviet Republics – and Western Europe – know that their well-being would be contingent on his good graces. Generally, he did so by issuing threats. However, during the coldest part of winter 2009, he went so far as to actually shut down the largest natural gas pipeline into Ukraine, whose government happened to be in arrears toward Gazprom. The result of the accounting discrepancy and fight for control of the pipeline, which serves countries to the west of Ukraine, was a three-week period during which millions of Ukrainians shivered in their dwellings and many surely died.

Later the same year, Putin was warning that the pipeline might be shut down again. Understanding the Russian leader's actions regarding energy and climate requires not just an understanding of any oligarch's thirst for power, but also the knowledge that he does not foresee unstoppable global warming or anything of the kind. He is not pretending. His best scientists, his top solar physicist among them, have told him, repeatedly, that the coming cold climate regime will be a time when conventional energy will be especially vital.

In the United States, meanwhile, on the heels of Al Gore's triumphant Academy Award for Best Documentary and Nobel Peace Prize, skeptical views on "climate change" were being belittled as fringe, corrupt, or worse. Despite centuries' worth of research into the subject, talking about sunspots, and any putative effect on Earth's climate, was clearly still a good way to raise eyebrows. In other words, a low-level Inquisition continued. NASA, as an example, found itself institutionally compelled to allude to the Sun-Earth climate connection and then to dismiss such talk as mumbo-jumbo, sometimes at the same press conference.

As NASA scientists watched the pace of the transition between the outgoing and incoming solar cycles with increasingly nervous eyes, a December 2007 press release underlined expectations for a powerful new cycle:

> Many forecasters believe Solar Cycle 24 will be big and intense. Peaking in 2011 or 2012, the cycle to come could have significant impacts on telecommunications, air traffic, power grids and GPS systems. (And don't forget the Northern Lights!) In this age of satellites and cell phones, the next solar cycle could make itself felt as never before.

An implicit plea for continued funding could hardly be mistaken. While James Hansen and his NASA climatology team depend for their high level of funding on the perception that manmade global warming is a threat to humanity's continued existence, the agency's solar physicists have less widely known threats upon which to pin their funding hopes. The two groups, meanwhile, stood to help each other, if they could convince Congress and the public, with their respective expertise, that only CO_2 could have produced the recent warming. A Sun that was not on the verge of a major shift in activity was one stone in the foundation of the argument. Just over a year later, the agency's jaunty tone had shifted to one that was more defensive about the now visibly lingering solar minimum. A new press release bore the headline "What's Wrong with the Sun? (Nothing)," and was laced with reasons why the gap between sunspot cycles wasn't unusual at all. "It does seem like it's taking a long time, but I think we're just forgetting how long a solar minimum can last," Hathaway said. Past lengthy minima had taken place before, the press released explained, helpfully adding: "Most researchers weren't even born then."

NASA's reputation in general, and Hathaway's in particular, meanwhile, were being harmed by the Sun's failure to cooperate, and Abdussamatov was beginning to look increasingly prescient. As the minimum continued, with sunspots coming in fewer numbers and all other indices declining as well, a press conference in October 2008 in which Hathaway did not take part tried to correct the public perception that the entire agency had been caught flat-footed. The Ulysses space mission launched in 1990 had been collecting solar data, a team of NASA physicists explained, that corroborated the fact that the current solar minimum was the deepest of the satellite era, with the weakest solar wind during the past 50-plus years. Among the mission scientists participating in the news conference was Nancy Crooker, who said the following:

The length and depth of the current minimum is fully within the norm of the last two hundred years, so it's not unusual in that regard. But it is unique in the

William Livingston carefully measures the size and strength of sunspots. It may be that when sunspots disappeared during the Maunder Minimum, they merely became too faint to discern. *A sunspot group, left; William Livinston, right.*

space age. We also know that the Sun entered an extended minimum during the second half of the 17th century, but that's not likely to happen. Just yesterday, there was a significant sunspot emerging, and that's a good sign that we're moving into the next solar cycle.

Agency anxiety that the solar influence on climate would be brought to light was discernible in Crooker's every word, and yet, of the panel of five scientists speaking, she was the most open to discussing the Sun-Earth climate connection. The sunspot to which she referred did not prove to be any kind of ramp-up point, however, as solar activity returned to its previous low levels within days.

Around the time of the press conference featuring his agency colleagues, Hathaway withdrew from the NASA Solar Cycle committee that he, until that point,

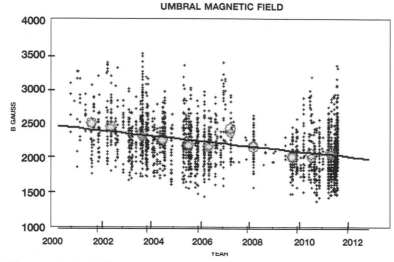

UMBRAL MAGNETIC FIELD

The graph, by William Livingston and Matthew Penn, of the National Solar Observatory, depicts the magnetic strength of sunspots versus time. Although sunspots can have magnetic strength less than 1500 Gauss, that is the lowest visible value. Extrapolating the linear trend forward, sunspots are predicted to become invisible by 2022.

. . . .

had chaired. The discussions, he said, had grown "too hot." Probably more significant, he moved away not only from his prediction of a strong Solar Cycle 24 but from the governing model of the Sun. "I have to admit that the last two months have been really disappointing," he said. "I have become increasingly worried about the conveyor belt and its role in the sunspot cycle and in the prediction of sunspot cycles."

In June 2009, after several more NASA press releases observing that the new solar cycle had yet to pick up meaningful momentum, a new optimistic release was put forth: "Mystery of the Missing Sunspots, Solved?" The new theory was that a "solar jet-stream" had taken an extra year moving into position to start generating sunspots. Although the critical new research was not done by a NASA team, the agency was keen to announce that all was well. The lead researcher, Frank Hill, of the National Solar Observatory, was on top of the world. "It is exciting to see," Hill said, "that just as this sluggish stream reaches the usual active latitude of twenty-two degrees, a year late, we finally begin to see new groups of sunspots emerging."

While spots were indeed beginning to show up with slightly greater frequency on the disk of the Sun, other major indices of solar activity remained low. Additionally, there was something curious about the new spots. They were vanishingly small, and unusually pale. The faint tones had been observed in increasing measure as long as fifteen years earlier by two other National Solar Observatory scientists, Matthew Penn and William Livingston. The two had attempted to get their observations of lower-contrast sunspots as well as analysis of the risk of a forthcoming Maunder-like minimum published, but with only partial success. Although the magazine *Science* had published several other of Livingston's papers, it declined to in this case, citing the short period of observation (about one solar cycle and a half). The *Astrophysical Journal* did accept two papers on these subjects from the pair, the first in 2006 and the second in 2007.

If Livingston and Penn's observations are correct and the trend of lighter sunspots continues, and late developments in June 2011 suggested that they were correct, then sunspots will become invisible by the year 2022, if not sooner. The magnetic phenomena that produces them would still be taking place, and the spots would still be there, they just wouldn't be detectable in the visible spectrum of light. This may be what happened during the Maunder Minimum, and signal a grand solar minimum. Unfortunately, unlike Abdussamatov, Livingston and Penn do not have their president's ear.

Although Barack Obama indicated when he became president that he would "restore science to its rightful place" in American society, to those who take the unknowns of solar influence on Earth's climate seriously, the opposite has been the case. The pain caused by the president's complicity in the silencing of skeptics is palpable among plenty of researchers who think that carbon dioxide's power has been grossly exaggerated. Meanwhile, the effort to quiet skeptics has been extended to several continents. Among the scientists outside the U.S. who are wringing their hands is the Norwegian physicist Pål Brekke. "We could be in for a surprise," Brekke stated in 2007. "It's possible that the Sun plays an even more central role in global warming than we have suspected. Anyone who claims that the debate is over and the conclusions are firm has a fundamentally unscientific approach to one of the most momentous issues of our time."

While the Sun is not the only player in determining atmospheric temperature, acknowledging the Sun's partial role together with changes in land use and ocean cycles would be a good start for the leader of the free world.

In time, lay people may become as familiar with cosmic rays as they are with carbon dioxide today. *Henrik Svensmark, Canary Islands, 2000, in a still photo from The Cloud Mystery, a film by Lars Oxfeldt Mortensen.*

9 Unsung Heroes

The skeptics, on the other hand, are harder to find. Many of them, I would discover, don't want to be found at all and try very hard not to appear to be dissenters. They have no wish to be called names in the press, or to lose their jobs, or to have their funding cut off as many deniers have.

– From The Deniers, by Lawrence Solomon

What if I could get the president to study climate a little bit more completely? What else would I tell him? One thing: climate scientists as courageous as Galileo are being smeared, in ugly fashion, the world over. For if you want to sign your own death certificate, in a career sense, publicly admitting your rejection of the manmade global warming movement's tenets is one good way to accomplish the task.

Part of this state of affairs has to do with the fact that a sick, triangular relationship has emerged among scientists, journalists, and politicians. The three groups, to the extent that they are separate at all, effectively dare one another to ever-more extreme statements until the ground of truth shifts so far that few can ascertain where it actually is. Christopher

Booker and Richard North have written on such fear-mongering in *Scared to Death*.

> [T]here are few things most newspaper or television editors relish more than the chance to take part in a full-blown scare drama, devoting acres of newsprint and hours of airtime to the possibility that some sinister threat, which could cause huge numbers of people to die, is on the way. In this their closest collaborators are those scientists for whom the lure of publicity or the need to justify funding have come to overrule their scientific objectivity.

For present-day climate Galileos, pursuing their climate research to its logical conclusions remains, to some extent, career suicide. That is because when confronted with the choice between sometimes-corrupting funding and truly independent research, with children to feed and mortgages to pay, the funding becomes the only choice.

Ironically, among the most pervasive myths concerning global warming is the one pitting David against Goliath, in which those touting the perils of climate change are cast as David and skeptics as the vicious giant. Observers are required to ignore the facts: Media, most scientists, and governments the world over have spent and received so much money on their version of events that they have collectively become Goliath. Observers must also ignore the reality that skeptic scientists maintain their intellectual

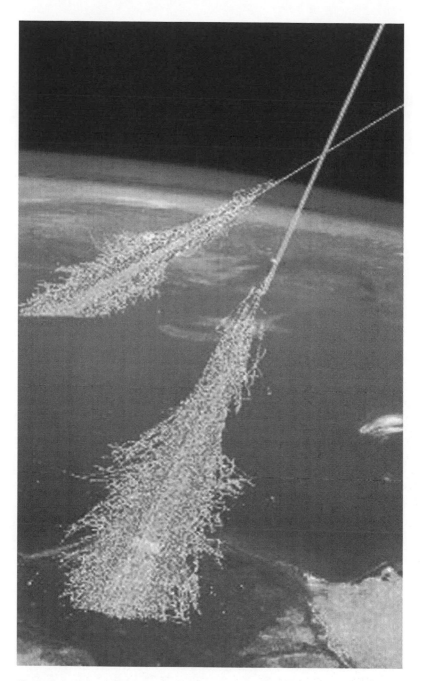

Cosmic rays, millions of which pass through Earth's atmosphere daily, appear increasingly likely to be an important component in the creation of clouds.

freedom at considerable risk. Funding for skeptics routinely dries up; tenure is denied them; ad hominem attacks of the most vicious variety are launched against them from the Ivory Towers of academia, from the studios of multi-billion dollar news organizations, and from the bully pulpit of governments the world over.

Big Oil, meanwhile, has been on its collective heels for a good ten years in the fight for public opinion in the face of such raw power. The oil companies spend next to nothing on scientific research performed by skeptics and even on public relations arguing that climate change is not a real problem. On the other hand, led by ExxonMobil, Big Oil has put well over a billion dollars toward funding *non-skeptical* research at the Climatic Research Unit, the University of California, and Stanford University, among many others. Shell, BP, and ExxonMobil have produced any number of slick advertisements touting their commitment to alternative energy that they more or less plainly don't believe in. Labeled "greenwashing" by environmentalists, the ads reveal the extent to which the rhetorical battle regarding climate change has already been won by the self-proclaimed underdogs.

The United Nations itself continues to spend billions of dollars a year on research, although the conclusions to be reached are arguably predetermined before the first computer-model run takes place. One of the leading atmospheric scientists in the world and a manmade global warming skeptic, Richard Lindzen

FLUCTUATIONS IN COSMIC RAY COUNTS

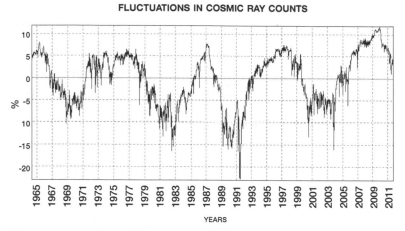

YEARS

The 11-year solar cycle's strongest influence on Earth's climate likely comes via cosmic rays. When the Sun is producing a lot of spots and other magnetic activity, the solar wind is strong and the number of cosmic rays goes down. Conversely, when the Sun is in a quiet phase, as it was in 2009, the number of cosmic rays goes up, in this case to the highest level ever recorded. *Cosmic rays, as measured at Oulu Station in Finland.*

• •

has said, "The consensus was reached before the research had even begun."

Adding up the *yearly* expenditures on global warming research of various kinds by the governments of the United States, the United Kingdom, France, Germany, Denmark, and the rest of the industrialized nations, one gets a figure upwards of 5 billion dollars. That is money that is supposed to be spent on hard science. As for public relations, Al Gore's foundation, the Alliance for Climate Protection, pledged in 2008 to spend $100 million a year for three years on advocacy work. If you have read a news story regarding climate change during this time, there is a very good chance that the article's origin was a press release from the Alliance for Climate Protection.

At times, in other words, Goliath's face is simply that

of Al Gore. The former vice-president and his team of speechwriters relentlessly distribute scare stories for public consumption, with Gore himself moving from one location to the next, and one media outlet to the next, usually by jet aircraft. With media access such as his, it is little wonder that his personal wealth has increased from $2 million on the day that he left office, in 2001, to more than $100 million today.

Unfortunately, there is no skeptic figure similar in stature to Gore, no expert questioning the scary narrative of climate change to whom the media give a free pass. More important, keeping skeptical pundits all but invisible, Gore crowds scientists from the stage. He is so successful that the public may be forgiven for imagining that skeptic scientists are few in number and low in professional standing. In too many cases, the media robotically sets a microphone in front of Gore rather than in front of a scientist. Such was the case in 2005, during the immediate aftermath of Hurricane Katrina. Said Gore at the time:

> There are scientific warnings now of another onrushing catastrophe. We were warned of an imminent attack by Al Qaeda; we didn't respond. We were warned the levees would break in New Orleans; we didn't respond. Now, the scientific community is warning us that the average hurricane will continue to get stronger because of global warming.

It is a fair bet that few members of the audience

listening to Gore's speech had any idea how controversial, and wrong, was the linking of any metric of hurricane power or frequency to presumed manmade global warming.

Expert after expert, most famously Christopher Landsea, had denounced such correlation as anti-science. In his resignation letter from the International Panel on Climate Change, Landsea, who works for the National Oceanic and Atmospheric Administration, wrote the following: "I personally cannot in good faith continue to contribute to a process that I view as both being motivated by pre-conceived agendas and being scientifically unsound." Unlike the former vice-president's remarks, however, Landsea's cry went unheard in the world media. He might just as well have never spoken. It is unconscionable that dissenting scientific voices are not being heard.

And hurricane research is not the only area in which scientists with skeptical views have been cast out of the cultural mainstream. Among the most compelling of these rogue scientists is Svensmark, whose ideas about clouds may one day make him the first Nobel laureate to be a climate skeptic. In the meantime, Svensmark, a physicist working for the Danish space agency, endures a seemingly endless stream of ad-hominem attacks, including by the head of the IPCC.

Generated by the explosions of ancient stars, cosmic rays fluctuate in inverse proportion to solar activity. The number of them passing through the atmosphere was at an all-time high from 2007 to 2009, and remains elevated in 2011. The reason is that the solar wind, the stream of particles leaving the Sun at all times, acts as a shield, deflecting cosmic rays away from the solar system, and away from Earth. During the 20th century grand solar maximum, the number of cosmic rays passing through Earth's atmosphere was relatively few. Conversely, when the Sun is in its quiet phase, as it has been recently, the solar wind goes slack, and cosmic ray counts go up. The arguments of Abdussamatov and like-minded solar physicists center around the notion that the oceans received unusual amounts of solar radiation due to the secondary effect of cosmic rays on clouds, during this decades-long stretch. They maintain, further, that the oceans are currently discharging this excess heat, even as a new cool regime begins to make its presence felt – in the form of record-cold winters in a variety of locations around the globe.

The fact that pure scientific research into such rays and their effects on clouds would come to be perceived as immoral, an "irresponsible" outcry against the narrative of global warming, is a poor reflection on the state of science in the early twenty-first century. What is beyond dispute is that the science of Svensmark and his fellow heretics is fascinating.

Svensmark's research is the most persuasive so far in establishing the link between cosmic rays and clouds. As far-out as this may sound, the scientific foundation upon which his theory has been constructed was actually begun more than 50 years ago. In 1959, Edward Ney published a paper in *Nature* entitled

"Cosmic Radiation and Weather." A young physicist who had been involved with the Manhattan Project, Ney had grown interested in cosmic rays in the years after the war. His paper linking the solar cycle, cosmic rays, and Earth's weather was simple and direct:

> The purpose of this communication is to point out the existence of a large tropospheric and stratospheric effect produced by the solar-cycle modulation of cosmic rays. Since there is some evidence for solar-cycle correlations in the weather, the phenomena described here should be considered in attempts to understand climatological effects of solar-cycle period.

Although at the time widespread knowledge of solar variability on the scale of multiple decades and multiple solar cycles was limited, Ney was zeroing in on the one form of solar variability that he knew existed: the 11-year sunspot cycle. Cosmic rays had been observed to increase during solar minima, and Ney said that, all other things being equal, such an increase in cosmic rays should lead to increased ionization of Earth's atmosphere, an increase that he set out to measure.

The existence of the solar wind, now known to regulate cosmic rays, was first brought to the world's attention in 1958 by Eugene Parker. As usual when someone departs from the scientific mainstream, Parker was belittled by his peers at the time. Two years later, however, the first of what would prove to be a series of observational confirmations came. Although it was a slow process, Parker's work eventually led to his receiving many of the highest honors in science.

Ney had worked throughout the 1950s on cloud chambers and cosmic ray detectors. The chambers, invented by C.T.R. Wilson in 1911, themselves have an interesting role in the history of climate research. Wilson demonstrated that cosmic rays left visible condensation tracks in the glass chamber. In the process, he implicitly answered the question: "How do you make a cloud?" He received the Nobel Prize for physics in 1929.

Ney, continuing Wilson's work, performed experiments for the United States Air Force, Army, and Navy, while a faculty member at the University of Minnesota. He became an expert in raising balloons to high altitudes in the process, setting a world-record in 1957 with a balloon that ascended to a height of 27 miles. In 1965, NASA astronauts took photographs for Ney to confirm his theories about zodiacal light, the dim glow in even the darkest night sky.

In fact, Ney was part of an evolving marriage between the world's various space programs and top-notch physicists. Although for a couple of decades the only two important players in the field of space research were the United States and the Soviet Union, European nations began creating their own agencies in the early 1970s. Among the most significant of these, it turned out, would be that of Denmark, two of whose physicists would come to have wide influence on the global warming debates of the 1990s and 2000s. These were Eigil Friis-Christensen and Henrik Svensmark.

The story of how Friis-Christensen became a global warming skeptic is an interesting one. In 1972, he went on a research expedition to his country's largest natural wonder – Greenland – and was fortunate enough while there to record an impressive solar storm. It seemed to Friis-Christensen, as he observed his instruments recording the unfolding storm, that such activity would have an effect on weather. Nearly two decades later, after much further research, Friis-Christensen published an article that was immediately controversial and influential: "Length of the Solar Cycle: An Indicator of Solar Activity Closely Associated With Climate." Although not positing a mechanism by which longer solar cycles would chill the Earth, the paper nonetheless drew on the work of Jack Eddy which associated a quiet Sun with colder earthly temperatures. Friis-Christensen saw that when the solar cycle persisted longer than 11 years, leading to a long and low solar minimum, temperatures on land in the Northern Hemisphere declined. When the solar cycle was shorter than 11 years, which it frequently was during the 20th century, temperatures went up.

In 1996, Henrik Svensmark began working for Friis-Christensen. Svensmark had begun examining cloud data – in search of a mechanism that would explain the link between longer solar cycles and declining temperatures on Earth that Friis-Christensen had found. Svensmark had gotten it into his mind that cosmic rays might be the agent responsible, and that it might have something to do with cloud cover. Russian scientists had also considered the possibility that clouds were regulated by cosmic rays, but had not managed to find anything conclusive. Once Svensmark obtained higher-quality cloud data, the relationship was clear. Svensmark (with Nigel Calder) wrote of the experience:

> The match was striking. Between 1984 and 1987 the Sun gradually became less stormy and more cosmic rays reached the Earth. Cloudiness over the oceans increased progressively by nearly 3 per cent. Then the cosmic rays declined till 1990 and the cloudiness decreased too, by 4 per cent.

As the research began to be fleshed out, Friis-Christensen told Svensmark that he was doing important, original work.

Soon, the two generated a paper that Friis-Christensen presented at a conference in Geneva. It so happened that at the same conference was Bert Bolin, the head of the IPCC. Asked by a Danish newspaper reporter what he thought of the research being performed by these two Danish scientists that linked cloud cover to cosmic rays, Bolin made a departure from normal scientific decorum, stating: "I find the move from this pair scientifically extremely naïve and irresponsible."

The effect of his remark was palpable. Svensmark and Friis-Christensen were shunned by their Danish colleagues, forced to scramble for funding that would normally be theirs for the asking, and criticized around the world. Svensmark was invited to a conference in his own hometown, for, as it turned out, a less-than-benign purpose. After dinner one evening, scientists literally shouted at Svensmark for having proposed his theory. Fortunately for him, the chairman of the International Commission on Clouds and Precipitation, Markku Kulmala, was at the event as well. Kulmala kept his peace until someone invited him to point out exactly which physical principles Svensmark's theory failed to observe. "It could be right," he said.

Despite Kulmala's expression of respect, the message, that the two Danes were up to no good, had been sent. Publication of their cosmic-ray based climate papers in the best journals would become fraught with difficulty. Journal editors were clearly concerned about what Svensmark and Friis-Christensen's work represented – a threat to global warming theory. State funding for the two men's work at the space agency even fell through, and one of Denmark's largest private donors to science in the country, the Carlsberg Foundation, ended up being enlisted to underwrite research.

In Switzerland, using a particle accelerator called CERN, an effort has been under way for a few years to test Svensmark's theory. Jasper Kirkby, a physicist working on the project, found that although requesting only a small amount of money for the work obstacles continually appeared. A reviewer of one of Kirkby's proposals came out and said that CERN was being used in a political theater that it should avoid at all costs. Eventually, approval was gained. The experiment, CLOUD, was conducted in 2010, and, interestingly, Kirkby chose to postpone the release of the results for two or three years. Nigel Calder interpreted the move as proof that the result corroborated Svensmark, but Kirkby remained reticent.

Svensmark and Calder wrote of the difficulties faced in trying to advance original work in such a political climate:

> Historians of science looking back on this little saga may wonder why both Kirkby in Geneva and Svensmark in Copenhagen had such a white-knuckle ride to get approval and funding for separate projects costing just a few million US dollars. The world was spending billions of dollars a year on climate research at the time. Further food for retrospection will be the assertions by opponents, including some very eminent scientists, that they knew that the results of the experiments would be negative.

As the battle to receive funding and time for the CLOUD experiment at CERN got going in earnest a decade ago, Svensmark and his team in Denmark proceeded with their own simpler set-up in a basement at the Danish National Space Center. The small experiment, known as SKY, showed that the key process of **nucleation** (or cloud-seeding) via cosmic rays was observable and measurable. It was real. The rays were found to liberate electrons that then initiated chain reactions involving microscopic droplets of sulfuric acid. A plane performing similar research over the Pacific had produced the same result from a slightly different perspective. What remains, basically, is for the results of the CLOUD experiment to be released – and for the real-world experiment involving the recent spike in cosmic rays to influence climate, or not.

The details of how the Sun and its modulating effect on cosmic rays affect Earth's climate remain to be sorted out, but dozens of scientists are working to accomplish exactly that. Among them are R. Giles Harrison and David Stephenson, who, in 2006, published a pro-cosmic-rays paper in *Proceedings of The Royal Society*. The pair point out that the Maunder Minimum remains the dominant conundrum for those who would discount a solar connection to climate on Earth. Eugene Parker, whose work regarding the solar wind proved to be significant in climatology, wrote the foreword to Calder and Svensmark's *The Chilling Stars*:

[I]t is our good fortune that, some years ago, Henrik Svensmark recognized the importance of cloud cover in the temperature control of planet Earth. Clouds are highly reflective to incoming sunlight. Svensmark also recognized that the individual water droplets that make up a cloud form mostly where ions have been created by passing cosmic ray particles, thereby tying cloud formation to the varying cosmic ray intensity. That is to say, cosmic rays control the powerful "cloud valve" that regulates the heating of Earth. There is an immense task ahead to quantify the effect, with some degree of urgency in view of the present global warming.

Curiously, the urgency has not immediately facilitated the acceptance and support for this research that one might reasonably expect. Global warming has become a political issue both in governments and in the scientific community. The scientific lines have been drawn by "eminent" scientists, and an important new idea is an unwelcome intruder. It upsets the established orthodoxy.

When it came to ostracism, of course, Parker knew of which he spoke. It was Parker who brought to Jack Eddy's attention the subject of Walter Maunder and his largely neglected papers. Eddy, who characteristically chose not to return any of the bitter scorn that his work attracted, died on June 10, 2009.

As several dozen scientists work to prove or disprove Svensmark's theories, his most important support has come from Israeli astrophysicist Nir Shaviv. Shaviv has said that until he looked into it personally he assumed manmade global warming to be a legitimate concern based on solid science. Once he began looking into it himself, however, his views changed. In 2002, Shaviv performed an analysis of fluctuations in cosmic rays due to the passage of the solar system through arms of the Milky Way galaxy.

His analysis confirmed that cosmic rays are a powerful climate driver:

> The variations in the cosmic-ray flux arising from our galactic journey are ten times larger than the variations due to solar activity, at the high cosmic-ray energies responsible for ionizing the lower atmosphere. If the Sun is responsible for variations in the global temperature of about 1 degree Celsius, the effect of the spiral-arm passages should be about 10 degrees. That is more than enough to change the Earth from a hothouse where temperate climates extend to the polar regions, to an icehouse with ice-caps on the poles, as is the case today.

In 2009, Shaviv published a new paper detailing a secondary effect of solar variation on the heat in the seas. The paper ("Using the oceans as a calorimeter to quantify the solar radiative forcing") is an elegant piece of work, deriving a picture of ocean heat from three independent sources, and showing a direct relationship to solar variability. Shaviv argues that there must be an amplification of this solar effect, on the order of five to seven times, to explain the magnitude of the changes in ocean heat content. His best candidate for that amplification mechanism is cosmic rays.

Despite the multiple research projects going on to verify or disprove Svensmark's work, he remains on the scientific margins, and in this he is far from alone. Some scientists have been cast out of the mainstream before ever making their way in. One of them is Steve McIntyre, the Canadian mathemetician who humbled NASA regarding its data a few years ago only to find himself at the center of the Climategate controversy in late 2009. While McIntyre holds no Ph.D., in the realm of statistical analysis he is formidable. In November 2009, the Climategate e-mails from the University of East Anglia's CRU showed just how seriously the climate-science elite take the Canadian. Thousands of e-mails from CRU were released to the press by a whistle-blower or computer hacker, and they revealed a world of bullying, backstabbing, data manipulation, and stonewalling. McIntyre's name was in 105 of the e-mails.

Meanwhile, Hansen, Schmidt, Jones, and Mann had long disparaged McIntyre in public. They said that he did not deserve to be treated with the respect of an academic, that he was meddlesome, and that he had nothing useful to say. The CRU e-mails, meanwhile, showed the lengths to which this clique ruling climatology had gone to block the publication of any rival ideas: everything from blackballing publications that dared to include a skeptic article, to forcing the resignation of an editor, to accepting inappropriate review privileges over rivals and close friends. "Peer-reviewed" had come to mean "friend-reviewed," in all too many cases, and had at the very least lost its connotation of intellectual detachment.

In the end, being cast out by the self-appointed elite among climate scientists may prove to be a badge of honor. In the meantime, with the global mean temperature failing to rise since 1998, and in the wake of what British journalists have typed as the largest scientific scandal of a generation, scientists with alternative explanations for the climate's behavior continue to have their work and their names besmirched.

AN
Exact and lively Mapp
or
REPRESENTATION
Of Booths and all the varieties of showes and
Humours upon the ICE on the River of
THAMES by LONDON
During that memorable Frost in the 35th yeare
of the Reigne of his Sacred Maty
King CHARLES the 2d
Anno Dni MDCLXXXIII.

With an Alphabetical Explanation of the
most remarkeable Figures

10 Euphoric Recall

For a long time it was thought that we moved into and out of ice ages gradually, over hundreds of thousands of years, but we now know that that has not been the case. Thanks to ice cores from Greenland we have a detailed record of climate for something over a hundred thousand years, and what is found there is not comforting. It shows that for most of its recent history Earth has been nothing like the stable and tranquil place that civilization has known, but rather has lurched violently between periods of warmth and brutal chill.

– Bill Bryson, A Brief History of Nearly Everything

With skeptical climate scientists struggling to make themselves heard over the roar of character assassination, there is a myth adhered to around the world: that the warming of the period from 1975 to 1998 was unusual. It was not. You do not have to go far back in time to see another period with the same kind

The bitter winters of the 17th century froze the Thames at London solidly enough for a carnival to be put on, as can be seen in this historical image. The winter of 2010-2011 was the third straight to be colder than normal in the United Kingdom. Whether the transition to colder winters will come to freeze the Thames again remains to be seen. *Frost fair, London, 1683.*

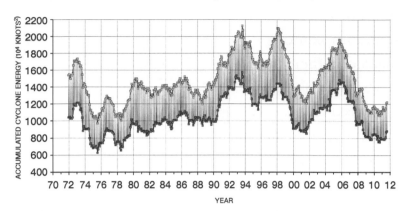

The rise of global warming alarmism in news stories and scientific publications coincided with a spike in tropical cyclone energy in the 1990s and mid-2000s. As of 2011, tropical activity has declined to historically low levels. *Global tropical accumulated cyclone energy since 1972.*

of warming – simply to the 35-year stretch from 1910 to 1945. Worldwide, temperatures, on average, rose just as much, and just as fast, during the early part of the 20th century as they did during the late 20th century: about a single degree Celsius. *In between, temperatures fell.* They began their decline at the precise moment that carbon dioxide was being emitted by human beings in significant quantities for the first time, during the post-WWII industrial boom.

And there is another, equally serious, myth afoot, one that draws nearly unlimited power from the euphoric recollection of weather. Central to the story is the idea that a few decades ago the climate system was more benign than today, that if enough money were spent in the right way climate would be benign again. A spin-off of the Garden of Eden narrative, the tale is designed to create longing for a time when the glaciers were larger, and the poles supposedly colder. Al Gore and his fellow story-tellers presuppose that if mankind could only go back and make a different decision when tempted by the serpent of progress, then glacial retreat, and climate change, would never have been visited upon us. It is an amazingly intoxicating fantasy. Daydreamers caught up by such visions are unlikely to know, however, that the actual living conditions of yesteryear were hazardous: longer winters followed by ice jams and grim floods, short growing seasons as well as non-existent growing seasons, followed by want and famine, and desperate rulers responding to the conditions around them with prolonged warfare over scant resources. These were "the good old days" of the Little Ice Age.

Meanwhile, climate has never been stable. Al Gore can parrot that weather was better-behaved until recently a thousand times, but that won't make it so. Both weather and climate change perpetually. Just about all of human history has taken place during the past 10,000 years or so, and it is no coincidence that our species has thrived during the warm interglacial period. Warm is better.

For, those of us living during the Holocene are in a gentle, life-giving nest, climatologically speaking. We are the beneficiaries of benevolent conditions such as very few generations of people living on Earth have enjoyed, with relatively abundant rainfall, long growing seasons, and lengthy periods of benign weather. Sometimes the "Holocene nest" gets rattled, it is true, for as long as a few centuries. But we are the lucky ones. We are the ones whom future generations, after the inevitable return to full Ice Age conditions, will look back upon with envy.

Our wonderful interglacial started in response to subtle eccentricities in Earth's orbit about 11,500 years ago. Greater and greater amounts of sunlight struck the planet at high latitudes (the most significant zone in regulating ice ages), melting occurred, relatively dark land and ocean were uncovered, and more sunlight was absorbed (rather than reflected by ice and snow). The Holocene had begun.

Almost immediately, things improved for most human beings, for this was summertime on an epic scale. Anyone who has ever lived close to nature knows how deeply the yearning for just a minor annual spring can be. After hundreds of thousands of years of using stone-age technology, of hunting great woolly beasts with spears, of migrating under extreme duress, of struggling to survive in grim cold, homo sapiens truly began to flourish with the arrival of our current climatic regime. Without the Holocene's gentleness, the technology that makes it possible for me to compose this sentence on my computer and for you to read it wherever you may be would not exist.

But for all the gentleness and stability of the Holocene, even the lesser warmings and coolings such as those of the past 150 years have had palpable effects on both weather and climate on both a local and a global scale. The temperature regime changes are registered in the popular press, such as in *The New York Times*, as we have already seen. One idea that surfaces in these jottings about climate again and again was that the Arctic was growing too cold or too warm.

> The Arctic Ocean is warming up, icebergs are growing scarcer and in some places the seals are finding the water too hot. Reports all point to a radical change in climate conditions and hitherto unheard-of temperatures in the Arctic zone. Expeditions report that scarcely any ice has been met with as far north as 81 degrees 29 minutes. Great masses of ice have been replaced by moraines of earth and stones, while at many points well-known glaciers have entirely disappeared.
> – The National Weather Service, as reported by the Associated Press, November 1922

And yet, thirty years after the ominous comment from the National Weather Service, the Arctic was in a cooling phase, which amplified Northern Hemisphere winters and led to the Ice Age articles, papers, and books during the mid-1970s. The colder Arctic from roughly 1945 to 1975 is one reason why, for instance, so many people who came of age during the 1950s and 1960s recall winter as having sharper teeth than it did, say, during the 1980s and 1990s. For the most part, the cold air, rushing down from the Arctic, *was* sharper when such people were younger. On the other hand, if you ask people who grew up a few decades earlier in the United States what it was like in the 1930s, they'll tell you that the weather was hotter. Period. And it was. In the United States, the warmest year on record is 1934, according to the Goddard Institute of Space Studies, and *several* of the hottest years were during the decade of the 1930s. The Dust Bowl on the American Plains and the historic migration it spawned were not created by cool temperatures.

And yet if the Dust Bowl were repeated today, 75 years after it took place, Al Gore and an army of videographers would surely present it as proof of human-induced climate change. The script would be easy to write: "*The suffering, the breathing-related illnesses, the loss of life – how much longer will some people keep their heads in the sand as industrial activity claims more victims?*" Too extreme? Not at all. People have been blaming natural phenomena on human behavior at least since "witches" were executed in classical Rome and in the rest of Europe until just a couple of centuries ago. More recent examples of such scapegoating abound. Gore, in the immediate aftermath of Hurricane Katrina in 2005, was walking on a well-beaten path, when he strove to blame the storm on certain members of humanity. "It is important to learn the lessons of what happens when scientific evidence and clear authoritative warnings are ignored," he said then, "in order to induce our leaders not to do it again and not to ignore the scientists again and not to leave us unprotected in the face of those threats that are facing us right now." The fact that powerful hurricanes had been slamming into

North America for thousands of years was no impediment to using the tragedy to reiterate the master narrative. The consciousness of every person with access to a television was brimming with images from New Orleans at the time.

Several points need to be made:

1. Photographs and video of extreme weather and climate are inherently powerful. Using such images to frighten non-experts is a tactic employed on a daily basis.

2. The variability of climate and weather are vastly underestimated by most members of the public.

3. During the past century, a time that is in no way remarkable by any geologic standard, the extremes of drought, wind, precipitation, and temperature have been at the same time catastrophic and utterly mundane.

4. As harsh as the worst of the Holocene interglacial has been, it doesn't approach the abrupt transitions from warm to cold and back again, from plentiful precipitation to almost none, from occasional gentle snowstorms for several winters to many years with mammoth, prolonged blizzards during the rest of the Ice Age that began, noticeably, three million years ago, and in earnest one million years ago. Violent and changeable though it has occasionally been, the Holocene has been a walk in

the park compared to what preceded it, and compared to what will certainly follow.

One hurricane flooding a coastal plain lying below sea level does nothing to prove the power of CO_2 to warm Earth's atmosphere. And yet the power of weather imagery remains. There is not one century, not one decade, not one year during humankind's presence on the planet that, if enough videotape were recorded, could not be edited into something terrifying.

A single aspect of Holocene variability comes to life when considering the Schnidejoch, an Alpine pass in Switzerland, written about by Henrik Svensmark and Nigel Calder. In 2003, a vacationing couple stumbled upon an archer's quiver that had recently melted free during the unusually warm weather that year. The couple turned the items over to grateful archeologists, who created a dig site and then kept it secret for two years. During that time, the scientists confirmed what various proxy measures of temperature had already indicated: that the world had been warming and cooling, dramatically, for thousands of years, a process that alternately buried the Alpine pass in snow and ice, and unburied it again. And it wasn't that a lone, intrepid soul had crossed the mountains using the pass once or twice in the past. The archeologists found hundreds of artifacts. An inn had even been constructed by the Romans to serve those using the pass during what we now call the Roman Warm Period. Although any number of climatologists and casual observers today bemoan melting like that of the Schnidejoch and insist that even though passes like it have opened in the past, *this time* it is a sign of something sinister.

Meanwhile, efforts by some scientists to restrict such climatic variation to a specific region of the world have failed, as dozens of peer-reviewed articles with evidence to the contrary attest. At the same time that glaciers grew and shrank in Europe, for example, they did the same thing in Chile, the Antarctic, and Greenland. As for the Schnidejoch pass in Switzerland, it did not open, this time around, *until 2003, after a 200-year recovery from the Little Ice Age.* That means that for more than 20 years, James Hansen and others were warning people of a warming that had yet to uncover an alpine pass that had been open, again and again, during the previous 6,000 years. The abundant evidence found by the archeologists as well as the nearby Roman inn suggest that the Schnidejoch remained open during summertime for lengthy periods of time, likely decades at a stretch, in the past. As of today, it has been open for seven years during summertime.

It is frankly difficult to believe that those promoting global warming have managed to link warmth with suffering and death. So far, during human history, warming has meant population growth and relative ease of living, not only as exemplified by the Holocene overall but by events within it. It is no coincidence that it was during the Holocene Optimum, from 6,000 B.C. to 4,000 B.C., that humanity began to practice agriculture, and to build important cities.

Another change during the Holocene Optimum was possibly that of rising sea levels. The Holocene Optimum high-water mark, known as a "highstand," was, according to many scientists, higher than where it is today. While not as impressive a highstand as that of the Eemian Interglacial 100,000 years ago, when sea level was 13 to 19 feet higher than today, it does underline the simple truth that the world's oceans have never maintained a constant level, and never will.

Yet a pervasive woe-is-me-ism has created an alternate reality in which only *we* have seen sea-level rise, only *we* have seen big storms, only *we* have endured drought, and only *we* have witnessed glacial melt. No matter how much evidence is compiled showing that such a view is absurd – and, further, laced with anti-Nature sentiment – some people cling to it as though to a holy relic.

Part of this panicky relation to the coast stems from the fact that our technology has created cities so vast and shiny that we demand that they possess an impossible and unnatural permanence. Unlike us, our forebears who endured rising sea levels during the beginning of the Holocene interglacial and highstands such as during the Eemian interglacial, did not have coastal cities to leave behind worth hundreds of billions of dollars. And indeed it sometimes seems that the projected horror of rising sea level (for it is possible that this interglacial will see a natural rise in sea level as high as during the Eemian) is the recognition of economic loss, above all. Our technology has allowed

Glaciers have been advancing and retreating by turns in the Alps throughout the Holocene, as recent archeological finds attest. *Climbers in Schnidejoch Pass, Switzerland, 2007.*

us to put a vast amount of wealth in harm's way, oblivious of the fact that building single houses, let alone beach towns, let alone major metropolitan areas, on the seacoast is an astonishingly optimistic thing to do.

One last detail from the last 12,000 years' worth of climate change: the northern forests of Siberia, now limited to an area ranging from 500 kilometers to 1,500 kilometers south of the Arctic Ocean, *extended all the way to the coast during the Holocene Optimum.* The trees tell their own story: today's climate is not particularly warm by the standards of the current interglacial.

Despite whatever risks it carries, warm is good. The current configuration of land and water, despite video trying to convince us otherwise, is benevolent. The Roman Warm Period from 200 B.C. to 535 A.D., a warm time like our own, was a period of abundant crops that benefited the empire. Conversely, the cold Dark Ages, from 535 A.D. to 900 A.D., saw the Nile River freeze, major cities abandoned, the fall of the Roman Empire, the collapse of the Mayas' civilization, and widespread pestilence and famine.

Conversely, the Medieval Warm Period, from 900 A.D. to 1300 A.D., was an era in which agriculture flourished, wealth increased, and dozens of lavish examples of Gothic architecture were created in Europe. A sign of the ease of living, compared to the Dark Ages, was the relative absence of major wars. Instead of such wasteful conflicts, you had Chaucer, Petrarch, courtly love, tapestry making, population explosion, abundance. One climate-driven historic event from the Medieval Warm Period that is difficult to fully digest is the Vikings' colonization of Greenland. The colony included the raising of sheep, the growing of crops, the building of a church, weddings, burials, visits to and from Iceland, and generally an elaborate, bustling outpost that lasted more than a century longer than the United States has existed so far.

The cold period from 1300 to 1800, on the other hand, saw plague, wars, crop failures, witch burnings, food riots, bloody revolutions, and the end of the Vikings' Greenland colony. Bigger glaciers, yes, but the human toll was high. Today, six and a half centuries later, the Greenland colonists are mostly buried in permafrost that has yet to melt. As for witch hunts, it is easy to forget how widespread the practice was and how bound up with famine. Of the hundreds of thousands of women executed as witches, crop failures were the most frequently cited crime requiring a penalty of public execution.

Colonialism was arguably a response to climate-induced scarcity, with the vast majority of expeditions heading to regions with warmer weather, and greater opportunities to grow crops. The canals of Holland were frozen winter after winter; the Thames at London was frozen solidly enough for ice fairs to be held on its surface. Then, after all this suffering and cold, there was good news at last: The Earth began to warm. For the Little Ice Age was followed by the Modern Warming of 1800 to 1998, and during this time population increased dramatically, technological and medical advances multiplied, and agriculture flourished. Although frost fairs on the Thames stopped being held, crop failures in Europe also become, by and large, things of the past.

How anyone could examine human history and see the advantages of keeping the temperature *lower* is a mystery. Agricultural production on the Canadian plains and Siberian steppes, both major sources of grain for the world's hungry mouths, are both at the northern limit of viability. If you could cool the Earth with the use of a thermostat dial, you would do so to the detriment of these farmers and the hundreds of millions of people that they feed. Again, when it comes to climate the good old days aren't always the good old days. Sometimes, the good old days are right now.

Another manner in which the Holocene's climatic variability actually favors those of us alive in the current day is precisely the rate of tropical-cyclone development. While some decades (and centuries) have had more frequent storms, former IPCC co-author Landsea has been joined by heavyweight scientists by the score who reject any link between increased carbon dioxide and increased hurricane activity. Perhaps the most famous of them is William Gray, whose yearly hurricane forecast is among the most important climate products. Gray fails to see any link between human industrial activity and climate change – most especially the number and intensity of tropical cyclones – and his research has been confirmed sufficiently that Al Gore no longer includes hurricane scare stories in his speeches.

Some of the best research into the rate of tropical cyclone development has been done by Ryan Maue at Florida State University. A 2009 study conducted by Maue concluded that tropical cyclone activity, plotted as a two-year running average, *was the lowest that it had been in 30 years*. Weather watchers in North America could be forgiven for finding the results surprising, for a couple of reasons. The first was that North Atlantic tropical cyclone numbers had been high during the decade of the 2000s. That fact, combined with relentless attention given to hurricanes by western media, especially American media, especially The Weather Channel, created a false sense of what was going on in the rest of the world. The relatively high level of Atlantic hurricane activity was connected to something called the **Atlantic Multidecadal Oscillation** – basically the

Atlantic version of the Pacific Decadal Oscillation. It had switched from cool to warm in 1995, and hurricane numbers had soared.

However, the North Atlantic is a tiny portion of the world ocean. Factoring in tropical cyclone activity from the Indian and Pacific Oceans, the recent *worldwide decrease* in activity has been steep, about 50 percent since 1998. And, as tropical cyclone activity is a fair metric of ocean warmth at the surface, the decline makes sense. The Met Office Hadley Centre's value for global sea surface mean temperature anomaly, meaning divergence from average, also peaked in 1998 at a value of 0.54 degrees Celsius. As of May 2009, the global ocean surface temperature was at 0.30, a significant 11-year decline. The fact that tropical cyclone development was on the downswing wasn't surprising. Either way, neither the recent lows nor recent highs of hurricane development are at all unusual for the Holocene. Instead, the storms are likely a reflection of standard interglacial conditions with a warm atmosphere, warm water in the tropics, and cold poles, and in this regard can be seen as an exception to the rule that the Holocene is gentle. Tragic though Hurricane Katrina was, it does not compare to the Galveston Hurricane of 1900 which killed 8,000 people, decades before the post-WWII rise in carbon dioxide emissions. The simple truth is that powerful ocean storms have occurred for as long as there have been oceans.

Another form of variability that bears examination is that of carbon dioxide itself. Although CO_2 in the atmosphere has increased from about 280 parts per million in 1800 to about 390 today, the modern value is still *many times smaller than what it has been for the great bulk of the planet's existence*. The sincere claim has been made by several skeptic scientists that plant life is starved for carbon dioxide. They point out, too, that the carbon dioxide added by man to the atmosphere comes from burning fossilized plants that sucked it out of an atmosphere *ten to twenty times more abundant in CO_2* than today. All of the fossil fuel – all of the coal, all of the natural gas, and all of the oil – plus all of the limestone on Earth was, previously, atmospheric carbon dioxide. The notion that a lot of carbon dioxide in the air is *unnatural*, unprecedented, or a negative of any kind is an opinion and not a fact. If you could dial down the world atmospheric level of carbon dioxide today, you would do so to the detriment of farmers and foresters *worldwide*.

It is not even established that carbon dioxide has varied during our own time strictly, or even primarily, due to human activity. As explained previously, degassing has a role in carbon dioxide's fluctuating levels. Here are the numbers for yearly carbon dioxide increase at Mauna Loa Observatory during the past 25 years:

Again, no one can tell you that the Super El Niño of 1997 was caused by human activity, but that year's warm seawater *certainly* caused the largest rise in carbon dioxide of the entire Mauna Loa record. Some have suggested that the 800-year lag in the relationship between temperature and carbon dioxide means that the Medieval Warm Period may be producing some portion of the current uptick in carbon dioxide through degasification.

So, before turning to the next chapter, let's go over a few things. Temperature has fluctuated, sometimes strongly, throughout the Holocene, with periods of warmth and cold alternating. The warm periods have been, by and large, easier to live through. The cold periods have been marked by famine, disease, witch hunts, and war.

Tropical cyclone development varies with time, and has been diminishing for at least fifteen years. The next time it rises, the global warming camp is likely to claim it for P.R. purposes, but that won't make it unusual.

The level of carbon dioxide has always fluctuated, but has not recently come close to its historic heights of hundreds of millions of years ago. Increases in carbon dioxide are part of nature, and are better understood as a response to heating rather than a cause.

In the meantime, remember: The good old days are right now.

One of the politicians who will forever be known for their efforts to put global warming at the center of public consciousness is the former vice-president of the United States. *Al Gore in Washington, 1994.*

11 A Broken
Moral Compass

Congressman, I have read all 648 pages of this bill. It took me two transcontinental flights on United Airlines to finish it.

– Al Gore testifying before Congress, April 24, 2009

As I've explained already, I was predisposed by birth and upbringing to concede the high moral ground to Al Gore and anybody else who told me that they were fighting on behalf of Mother Earth.

I was wrong.

The belief in the West that human beings are a scourge on the face of the planet has been discredited several times, and yet it remains remarkably widespread. Not only is the early twenty-first century an especially good time to be alive, in terms of climate, it is also an especially good time to be alive in terms of the environments in which people actually live. In our era, a higher percentage of humankind benefits from clean water, clean air, modern plumbing, and safe and reliable heating sources than ever before, and increasingly so. That said, I am proud that my country's Environmental

Protection Agency and Congress have acted in the past two generations to improve the quality of America's air and water, both of which were headed in bad directions. It is nice that our rivers no longer catch fire, for one thing. However, the argument that wealth, coupled with reasonable regulation, leads to a cleaner environment, is borne out by history, including and especially that of the United States. Working to continue this progress is an appropriate use of public funds and remains a priority for me personally.

Nonetheless, the moral superiority of Al Gore is not real. I recognize that he aspires to be moral. I recognize that morality is indeed a governing motivator for him. But as one of the leaders of the drive to create a new carbon trading market and to promote carbon trading among nations, he appears to have developed a peculiar form of moral blindness, nonetheless. If he were not such a powerful figure in the world's political landscape, I would not consider this to be any of my business, but he is and I do.

Gore bristles when accusations of being driven by material profit are spoken in his presence. Such was the case in April 2009, when he appeared before Congress to support the historic cap-and-trade bill being considered. Republican Representative Marsha Blackburn of Tennessee asked whether Gore stood to profit personally from the legislation. Amid sighs and condescending laughs, the former vice-president metaphorically rapped his fellow Tennessean on the knuckles. "Congresswoman," he said, "if you believe

Turning food appropriate for human consumption into automobile fuel is problematic. It has direct consequences and indirect consequences, most of them disastrous. *The unnecessary turning away from fossil fuels to soybeans, left, has made tempeh, wrapped in banana leaves, center, more expensive in the Third World. Ethanol from corn, right, grows from bad economics and bad science into bad public policy.*

that the reason I have been working on this issue for thirty years is because of greed, you don't know me."

Was the former vice-president a partner of Kleiner-Perkins? Blackburn persisted. He was. Was he aware that the company had invested a billion dollars in enterprises that were poised to profit from cap-and-trade legislation? He was. At this point, the former vice-president had a question for Rep. Blackburn: "Do you think there's something wrong with being active in business in this country?" Gore also insisted, vehemently, that "every penny" from his book, his movie, and his corporate work in the emerging carbon market was put into his nonprofit educational foundation, the Alliance for Climate Protection. That may or may not be literally the case, but it may not matter, in the end. Sitting on Google's board, as he does, Gore stands to profit from legislation that favors Google's ideas about what a "green" economy should look like. And, one interesting question remained after the fiery interchange between the representative and Gore: How much personal wealth had he accumulated of late?

Estimated to be worth two million dollars when he and Bill Clinton left office, by 2007 Gore's real-estate, investments, other holdings, and cash had grown in value to 100 million dollars. A *Fast Company* article by Ellen McGirt ("Al Gore's $100 Million Makeover") mentioned what everyone already knew: that the ex-vice-president had become an adviser to Google and a board member of Apple. What many didn't know was that he had made a total of 36 million dollars from the two companies. With more millions coming in from speaking fees and book and movie profits, Gore had become an extremely wealthy man, altruistic or not.

According to Bloomberg News, Gore was at any rate aggressively re-investing the money in hedge funds and a variety of private partnerships. In the meantime, a man visibly upset by the emission of carbon dioxide was literally circling the globe by jet, repeatedly, to spread the news. His carbon footprint in the state of Tennessee was famously gigantic, with a yearly electric bill twenty times that of the average U.S. home. In 2008, he purchased a houseboat for use on Tennessee's Center Hill Lake. Wasn't buying a fuel-hungry recreational craft a risky move for someone on a mission to dial down CO_2 emissions? Ah, but the former vice-president was veritably lying in wait for just such a criticism. His response? *The boat ran on bio-diesel.* Or at least it could – theoretically. In the end, it turns out that no one on Center Hill Lake actually sells the stuff.

The moral righteousness of bio-diesel is imaginary, by the way, on more than one level. The single biggest reason is that running a vehicle on food when hundreds of millions of people on Earth remain starving or malnourished is an ugly piece of moral calculus. It is not just a question of food, per se, but also of protein. For the global soy market has come to be dominated by American farmers during the past two decades, with soy growers in southeast Asia simply unable to keep up with the low prices that their subsidized American competitors can accept.

In the meantime, soy beans in the millions of metric tons get turned into motor fuel. The consequences for people living in Indonesia and the rest of the Third World have been dire, as Associated Press writer Michael Case wrote in September 2008:

> The cost of soy is spreading hunger on the country's main island of Java, where millions of poor and working-class families depend on tofu and tempeh every day. It is also devastating an entire local industry based on soy products. Hundreds of factories have closed, thousands of people have taken to the streets to protest soy prices and at least one soy vendor killed himself after falling into debt.
>
> About 20 percent of soy now goes to make biodiesel in the U.S., up from almost nothing three years ago...

Converting soybeans to motor fuel sounds like a good way of increasing energy independence. But what right-thinking person would have chosen to expand the use of soy-based biodiesel, knowing in advance the effect it would have on those struggling to provide their families with adequate protein? Using biodiesel for recreational purposes, especially, becomes the kind of thing that is hard to point to with pride.

Now, in addition to advocating an ill-considered transition away from petroleum-based motor fuel to agriculturally derived motor fuel, Gore is leading a movement to create a "new economy" based on the buying and selling of carbon credits. Apart from the fact that the carbon dioxide generated by humankind likely has little impact on the ocean-atmosphere system, the cap-and-trade gambit is another idea that is likely to have any number of unintended consequences.

On the geopolitical level, when a poor nation with no electric grid contemplates pulling its economy into the twenty-first century, the decision to outlay capital is daunting enough, even without the added burden of paying for the emission of carbon under any international agreements. On the business level, the cap-and-trade system being contemplated by the United States Congress is tremendously unwieldy, rife with opportunities for corruption, probably impossible to enforce, and unlikely to generate anything in the way of measurably lower temperatures thirty years from now – or ever. That is why James Hansen, of all people, considers cap-and-trade to be offensive, arguing that it achieves far too small of a reduction in greenhouse gases. In the end, cap-and-trade may never be passed, but the intention underlying it has already been given real teeth by the Obama Administration. This is because the EPA has declared CO_2 to be a dangerous pollutant and has been granted sweeping new powers with which to regulate it.

As the U.S. lurches forward into the realm of cap-and-trade, even the relatively tame provisions of Kyoto, and those in play in Europe, create real hardship for everyday people. James Hansen's idea to tax carbon dioxide is horrific, but it at least puts the reality of the extreme environmentalists' position right in front of the public. What cap-and-trade will achieve, and has already achieved in Europe to a significant extent, is simply higher energy prices – and a few wealthy carbon traders.

The underlying value of traded carbon is zero. The financial meltdown of 2008 was surely a small-scale affair compared to what might very well unfold after burdening the western economic system with the trading of carbon. Such a collapse would have one familiar parent to those examining the issue of climate: computer models. As with global circulation models and the mortgage-based-securities models, the carbon-trading computer models reflect the biases of their programmers, who would be distancing real wealth from human productivity in a way that can only be described as dangerous.

With hidden taxes appearing in the energy system, and thus within the entire U.S. economy, it will not be the privileged elites who will suffer the most when the "new economy," like the one that melted down in 2008, comes crashing down. It will be those on fixed incomes who pay, particularly those living in northern latitudes, who in many cases will have to make the choice whether to pay heating bills, medical bills, or grocery bills.

This grim financial reality is already a mainstay in Britain, where energy costs for consumers have nearly doubled in the past decade in response to the U.K.'s self-destructive commitment to diminishing its carbon footprint. An energized British environmental movement has stymied efforts to construct even a single new coal-fired power plant, despite the country's considerable coal reserves. The U.K. has spent hefty sums, meanwhile, ramping up construction of wind farms. Unfortunately, not even wind is morally straightforward, and it is most certainly not economically straightforward. Because the wind itself is so unreliable, wind farms cannot be tied into electric grids without also building additional, redundant, conventional plants, be they nuclear or fossil-fuel fired. In order to keep the entire grid functioning, the conventional plants have to have their turbines running for the many occasions when the wind stops blowing, as it usually does without warning. Electric grids are neither simple nor small things. They are, rather, remarkable engineering achievements that can be compared to the tissue of a living body. Just like our own tissue, they get lifeblood where it needs to go, seemingly without effort. But the effortlessness is a fiction. Switching feel-good but unreliable energy sources such as wind power into such a complicated system is a dicey proposition.

The winter of 2009 in the United Kingdom provides a good example. The coldest temperatures in 30 years were accompanied by a near complete lack of wind. When power was needed most, it was least available. The problem with building redundant power plants is that they cost as much to build as ones that run full-time. The fact that they are supposed to be used less does not diminish their initial price tag by a penny. In the end, the future of wind power in the United Kingdom is anything but clear. It was announced in April 2009 that Britain's only wind-turbine manufacturing plant would be closed.

Three other countries in Europe have made considerable efforts to convert to wind power. They are: Germany, Spain, and Denmark. Denmark is the most wind-powered of all, and it is far from a success story. Although at times the country's wind turbines generate a percentage of its consumption somewhere in the teens, generally speaking the power is produced at times when the power is not needed and it is sold for little or nothing to its neighbors. At other times, and frequently, Denmark must import power. For the most part, the purchased power comes from Sweden (mostly nuclear) and Norway (mostly hydroelectric), as Christopher Horner and others have reported. So, while Denmark takes considerable pride in its conversion to wind power, and takes pains to point out how ugly a thing nuclear power is, when the wind goes slack, it becomes a nuclear power consumer, on the down low. So much for the moral high ground, and so much for the reliability of wind.

Similarly, the most ambitious plan for wind-power production in the United States, one envisioned by oilman T. Boone Pickens, was scuttled in July 2009. Pickens had planned a vast wind farm in the northwestern quadrant of Texas, spending millions of dollars nationwide on advertising to develop support for the idea and two billion dollars on turbines manufactured by General Electric. Moving the power from its remote location to the electric grid proved to be too daunting of an obstacle, even for the hard-boiled Pickens, who in December 2010 began trying to sell his turbines to Canada. The Texas wind farm may or may not come back to life when all is said and done, but no one should be under any illusion what it will represent if it ever comes online. The redundant systems will always have to be there.

In the fight to be (and appear) moral, electric and hybrid vehicles have become important symbols and big sellers, compared to just a decade ago. But where do reality and perception meet with cars like these? While it may be very nice to see no exhaust coming from one's own car, plugging the vehicle into the electric grid does not eliminate fossil-fuel pollution – far from it. In fact, the greater the distance that electricity has to travel to get to a consumer's garage, the more energy is lost along the way. Given that coal-fired electric plants are going to be a necessary evil for another generation or two, at least, using electricity from the grid for private automobiles *increases* the proportion of energy coming from this relatively dirty source. On top of that, the batteries used in both electric vehicles and hybrids are enormous. The bad news here is that there has yet to be a battery produced that does not have toxicity issues. Buying a hybrid, to a large extent, means increasing the battery toxicity problem that humanity will eventually have to solve. It *feels* moral to drive an electric car, but it probably isn't.

Another way that energy perception and reality diverge has to do with prioritization. As various nations and industries strain not to use carbon-based fuels – even when they remain the best fuel sources at their disposal – it means diverting the same money away from dozens of *real* environmental and social problems begging for resources. The fact that any second or even third-tier scientist can easily obtain funding for a study to establish that squirrels have changed their mating habits in Buckinghamshire due to manmade global warming has a very serious moral consequence. Why not spend the same $100,000 for that study on a Superfund toxic cleanup site, a scrubber for a coal-fired electric plant smokestack, or clean water for a community in the Third World? Bjorn Lomborg has written extensively about such prioritizing of problems. The quote below comes from his provocative book *Cool It*:

Malnutrition kills almost four million people each year. Perhaps even more dramatically, it affects more than half the world's population, by damaging eyesight, lowering IQ, reducing development, and restricting human productivity. Investing $12 billion could probably halve the incidence and death rate, with each dollar doing more than thirty dollars' worth of social good.

It is hard not to overhear conversations about global warming in public places throughout the west. But malnutrition? It has nearly fallen off the map – that's how successful public relations blitzes like those managed by Gore have been.

Spending vast sums to reduce the carbon footprint of the industrialized west, as India and China continue to grow theirs, simply will not diminish human suffering. The expenditures may or may not reduce the global mean temperature by even a tenth of a degree Celsius. Spending billions on mercury capture, though, and other real-world environmental problems, makes sense. The most effective mercury-capture systems in coal-fired smokestacks can yield as much as 80 or 90 percent reduction in emissions, according to the EPA. On the other hand, those neat little light bulbs that Al Gore has been promoting? Unlike conventional bulbs, they are filled with mercury. Although reasonable people can disagree about whether the gains in efficiency offered by the bulbs justify the mercury in them, there is no capturing the bulbs' poison once they make their way to a landfill.

No one, it must be remembered, has ever died from

a cancer caused by carbon dioxide. But there is a long list of real toxins being emitted into the environment every day, and an equally long list of other toxins that are already there, all of them requiring cleanup. That Al Gore should have distracted so many from the real problems facing the world does not seem very moral at all.

Finally, I want to examine the moral component of meteorological journalism. As I mentioned near the beginning of this book, I used to be an *avid* watcher of The Weather Channel. For a good couple of decades, the network was not only an important component for the fledgling cable industry, but an excellent source of information about current weather and climate, as well about atmospheric science itself. An interesting thing took place during the 1990s, though. Weather Channel viewership was found to spike during hurricanes, and not merely among viewers in areas that could be affected by the individual storm being discussed. A lot of people evidently loved watching the progress of tropical storms, the stronger the better. Hurricanes became, over time, a revenue producer for the network. Experts were hired and given regular on-air time, and hurricane segments were given their own titles, their own graphics, and their own music. People loved it.

Much of this was quite innocuous, and arguably inevitable. Hurricanes are indeed interesting, and for a period of about 15 years it was widely believed, even by many scientists, that manmade global warming was ramping up the number, intensity, and duration of storms. In the last few years, however, links between recent atmospheric warming and hurricane activity, as we have seen, have been reconsidered.

In the meantime, though, the false link had lodged in the popular imagination, and The Weather Channel was more or less avidly exploiting it. The network's presenters didn't overtly come out and say that individual storms were generated by tailpipe and smokestack emissions, but they didn't really have to at this point. The misconception was so pervasive and so widespread that merely trumpeting the "unusual" power of the storms themselves sufficed. In the meantime, the network slowly upped its on-air mentions of the phenomenon of global warming during the daily program cycle and eventually devoted a new segment to the phenomenon known as "Forecast Earth."

Video alarmism regarding atmospheric phenomena is, perhaps, to be expected by a network like The Weather Channel. After all, it is hardly alone. The major cable news networks routinely send meteorologists and other reporters into the path of hurricanes, so that they can be seen amid the rising waters, gusting winds, and torrential rains.

On the other hand, the tranquil weather being experienced by most people around the globe at any given time goes ignored and unvideotaped. Again, one can understand why this would be so. In the newspaper business, and other journalistic domains as well, fires are of note. Non-fires aren't. Fair enough. But something very insidious has taken place. The selling of weather disasters as entertainment has led to a state in which big business stands to gain handsomely from the perception that the planet has gone meteorologically mad. Specifically, General Electric stands to profit. When in 2008 NBC (owned by General Electric) purchased The Weather Channel, an interesting thing took place: the largest domestic producer of wind turbines became the owner of the best-positioned purveyor of images of destructive weather. The same year, NBC's Today Show continued its longstanding practice of "showing" the great destruction to the ocean-atmosphere system caused by manmade global warming, with story after story: fires, floods, melting Kilimanjaro, you name it. The rest of NBC News, and the Weather Channel, meanwhile, keep the same pieces of videotape on nearly infinite repeat.

Summing up: Wind turbines do not deliver reliable electric power; the ocean-atmosphere system is not broken; scaring people needlessly isn't nice – and it distracts them from the *actual* environmental problems surrounding them.

There are many reasons why temperatures may go down during the next few decades. One possibility is the veiling of the Sun by volcanic eruptions. *Eyjafjallajökull Volcano, Iceland, 2010.*

12 Don't Sell Your Coat

Ladies and gentlemen, the warnings about global warming have been extremely clear for a long time. We are facing a global climate crisis. It is deepening. We are entering a period of consequences.

– Al Gore

Since businesses, governments, and individuals around the world shouldn't be preparing for catastrophic global warming, as I have explained thus far, then what should they be preparing for? One answer, perhaps, is a period of global cooling.

There are at least five reasons why preparing for mild cooling in the next few decades would be a reasonable thing for governments to do:

1. **Homeostasis:** This is the property by which an entity or organism tends toward stability. In the case of temperature, our planet has it in spades. While analyses of global mean temperature records, shown as anomalies in tenths of a degree Celsius, look like a terrifying roller coaster ride on a graph, absolute temperatures on the same graph look nearly level. Even the planet's trip into and

227

out of the Medieval Warm Period and the Little Ice Age, when looked at on an absolute-temperature graph, look unimpressive. That's because, for all of the drama of a change of a degree or two, the Earth's atmosphere stays remarkably close to an average annual temperature of 287 Kelvins (14 degrees Celsius, or 58 degrees Fahrenheit). The mere fact that people think they know what climate "should" be is indicative of Earth's homeostasis. For all the ups and downs, and all their effects, there is a remarkable degree of temperature stability during all of human history.

The nature of homeostasis says that if the last 30 years represent the pendulum swinging toward warmth, then the pendulum should, logically, now swing toward cool. Any number of negative feedbacks in the climate system could yield this kind of cooling, including and especially increases in evaporation, cloudiness, and precipitation.

2. **The Pacific Decadal Oscillation (PDO) turning negative**. The last time this happened, the ocean-atmosphere system cooled for around 30 years.

3. **The Atlantic Multidecadal Oscillation (AMO):** Some argue that this has as large an effect as its Pacific cousin on the global mean temperature, or at the very least upon weather and climate in the Northern Hemisphere. When the AMO switches to negative sometime around 2015 and is in phase

with the PDO for approximately 15 years, the cooling effect may be pronounced.

4. **The grand solar minimum** predicted by Abdussamatov and others. Whether the cooling shown in the paleoclimate record resulted primarily from secondary effects, such as cosmic-ray created increases in clouds, an unidentified effect in the oceans themselves, or some combination of these with the decline in total solar irradiance, if we witness a grand solar minimum ourselves, it is likely that we will also witness cooling.

5. **Explosive volcanic eruptions**: By shooting sulphur dioxide and other fine particles high into the atmosphere volcanoes diminish the amount of sunlight reaching the surface of the Earth. The process is one form of what's known as global dimming. While the majority of such events have relatively brief cooling effects on the atmosphere ranging from a year to a couple of years, larger eruptions have produced longer-lasting cooling. NASA scientists and others have explored the link between solar minima and increased volcanic activity.

If cooling comes, there is every chance that it will be missed by many observers, at least initially. People who rely on the media to learn of cold weather around the globe may be forgiven for imagining that it has disappeared. It has not. Indeed, from the moment the

Pacific Decadal Oscillation switched to its cold phase in 2006, Earth's climate has been dishing up some destructive winters, producing, among other things, snow in Baghdad for the first time in 100 years. The winter of 2007-2008 was the coldest and snowiest in China in 30 years, killing an untold number of people. Even state-run media admitted that hundreds perished. The Pacific Northwest saw its coldest and snowiest winter in decades, with annual snowfall records exceeded in two successive years in Spokane, Washington. There were all-time records for cold and snow in the upper Midwest during 2007-2008 as well, with dozens of municipalities running out of road salt and having to purchase more at a premium as winter dragged on.

During June, July, and August 2008, the longest and best winter ski season in a generation was enjoyed in both Australia and New Zealand. Simultaneously, the opening of mountain passes through the northern Rockies was delayed after the season's unusually heavy snow. Another tidbit: an October snowstorm in London for the first time since 1922.

Parts of North Dakota received their greatest-ever snowfall in 2009 as well, with some locations receiving more than five times the average snowfall. There was the biggest snowstorm in Las Vegas in the past 30 years. Houston received its heaviest snowfall in memory, as well as one of its earliest snowstorms in history, which also gave many parts of Louisiana unusually early and unusually deep snow.

So much snow fell in Oslo, Norway, that there was

no place to put it within the city limits. Street cleaners ended up pushing mountains of white into the harbor. On the other side of the Atlantic, there was a very cold and snowy winter in New England as well. In January, Maine set its all-time record low: -50° Fahrenheit. Snow fell in the United Arab Emirates the same month for the second time in recorded history.

As summer 2009 started in the Northern Hemisphere, the snow came early once again to Australian and New Zealander ski resorts. The same month, there was the first-ever July frost on Prince Edward Island. New York City failed to hit 85° Fahrenheit during the month of June for the first time since 1916. Phoenix, Arizona, went two weeks straight the same month without touching 100°, the first time since 1913 (a significant solar minimum occurred that year).

Winter 2009-2010 saw the continuation of extreme wintry weather around the Northern Hemisphere, with *the most extensive snow cover ever measured.* Record snow and cold stretched from Dallas, Texas, to Washington, D.C., to the United Kingdom, to Scandinavia, Central Europe, Eastern Europe, and nearly all of Asia.

Most of the cold-weather events went unreported or under-reported, or, in those rare instances when they were given sustained attention, they were laid at the feet of human-caused climate change. Such no-matter-what-happens-it's-human-caused-climate-change thinking is something for which Al Gore is deservedly famous, and took place again during winter 2009-'10. Such pronouncements are part of a wider pattern among those on his side of the climate wars: One heat wave is proof of global warming, and seven extreme winter weather events are proof of damage to the climate system. In this way, all phenomena support the mythology.

Beyond confusing those who would like to understand the climate system, this form of bias is actually dangerous. Snow and cold are serious. Excess heat causes thousands of deaths worldwide each year, but cold kills far more. In 1999, Friends of the Earth estimated that 50,000 people had died in the United Kingdom alone for lack of funds to pay their home heating bills. That was before the doubling of British energy prices to fulfill the green agenda.

The meteorological bias is also why the blogsosphere has come to play a vital role for those fascinated by weather and climate. Sites such as wattsupwiththat. com raise the hackles of global-warming alarmists simply by reporting weather news that the mainstream media ignores.

One ironic emblem of the spate of cold weather during the last three years was President Obama's forced departure in December 2009 from the IPCC's climate conference in Copenhagen in order to be able to touch down in Washington, D.C., before thickening snowfall made landing Air Force One impossible. *The New York Times*' initial article about the president's change in plans mentioned that it was a "snowstorm" forcing the hasty departure; later versions of the article changed the word simply to "storm."

Mary Shelley, author of *Frankenstein*, penned her iconic vision of bitter cold during the volcano-induced Year Without a Summer, 1816. *The author, as painted by Reginald Easton.*

As recently as ten years ago, the U.K. newspaper *The Independent* warned its readers with the headline: "Snowfalls are now just a thing of the past." Meanwhile, municipalities throughout the United Kingdom have struggled to keep up with ice- and snow-covered roadways for the last two winters. In satellite views, the entire British Isles appeared as a ragged-edged cake covered with white frosting.

As for the next few decades, what if a volcano or two should erupt on a serious scale? Surprisingly few people are aware of how direct the impact of volcanic eruptions has been on even fairly recent climate, but volcanic eruptions cooling the atmosphere are actually quite common. Among the most famous of these was the eruption of Indonesia's Mt. Tambora in 1815, causing the "Year Without a Summer" in 1816. Summering on

Lake Geneva in Switzerland, Mary Shelley, Percy Bysshe Shelley, Lord Byron, and other prominent writers were forced indoors by cold, foul weather. The group held a famous competition, holed up in their villa, to generate the most eerie tale that each of them could. Two astonishing creations emerged: Mary Shelley's *Frankenstein*, begun at the villa when she was but 18, and John William Polidori's *The Vampyre*.

While the dark and stormy weather on Lake Geneva proved fertile ground for these writers, the effects of the Tambora eruption were grim around the world. Rivers that had not frozen in years suddenly did so and stayed that way longer than any time in memory; frosts came in summer; and, most important, crops failed. Famine was widespread – in India, China, Europe, North America – and many perished.

The sudden cold was anything but unprecedented, which, given the number of active volcanoes on the planet – thousands – is hardly surprising. Volcanoes' destructive capacity can be more significant than what was witnessed in 1816. Dark Ages cooling, as a case in point, was likely launched by a catastrophic eruption of Krakatoa in 535. Tree-ring data shows sudden widespread cold for the years 536 and 537; the warmth of the Roman Climatic Optimum would not be seen again for 500 years.

Less dramatic, but nonetheless significant, cool-weather-inducing volcanic eruptions of the last century and a half occurred at Krakatoa in 1883, El Chichon in 1982, and Mt. Pinatubo in 1991. Volcanoes are, simply put, part of the ocean-atmosphere system, a wildcard that cannot be removed from the table. People lose track of a simple fact in the climate debates: cold is harmful to human health (and to the well-being of most life forms). Mary Shelley, shivering in the effects of a volcanic eruption on the other side of the world two centuries ago, had occasion to think about the nature of cold and built it into her masterpiece, *Frankenstein*. In the lines that follow, Dr. Frankenstein tracks his creation into the frigid north:

> As I still pursued my journey to the northward, the snows thickened and the cold increased in a degree almost too severe to support. The peasants were shut up in their hovels, and only a few of the most hardy ventured forth to seize the animals whom starvation had forced from their hiding places to seek for prey. The rivers were covered with ice and no fish could be procured; and thus I was cut off from my chief article of maintenance.

People forget: cold and ice are not our friends.

The other climate wildcard, the big one, is the end of the Holocene itself. Some believe that we are overdue to descend into the "normal" that is our current Ice Age. Others believe that the Holocene is not "scheduled" to end for few thousand more years. If the conclusion of the Holocene takes place anytime soon, perhaps people's views of the significance of manmade carbon dioxide will alter.

Of more immediate relevance, cycles in the world's largest bodies of water are likely to push the global mean temperature gently downward for the next few decades. This cooling may be amplified by a continued downturn in solar activity, via cooling "Svensmark" clouds. In retrospect, the twenty years or so that humanity has spent fearing the effects of elevated carbon dioxide on the ocean-atmosphere system will appear as an embarrassing episode in the history of science.

Nature's fury did not start when The Weather Channel opened for business. *Flood victims are paddled to safety; Dayton, Ohio, 1913.*

Bibliography

· · · · · · · · · · · · ·

Chapter 1: A Map for the Climate Battleground

Ahn, Jinho, et al., "CO_2 diffusion in polar ice: observations from naturally formed CO_2 spikes in the Siple Dome (Antarctica) ice core," *Journal of Glaciology*, Volume 54, Number 187, 2008

Bhattacharyya, A., et al., "The Roman and Medieval Warm Periods at Paradise Lake, Northwestern Himalaya," *Current Science*, Volume 93, 2007

Cai, W., and Cowan, T., "La Niña Modoki impacts Australia autumn rainfall variability," *Geophysical Research Letters*, Volume 36, 2009

Ge, Q.-S., et al., "Temperature variation through 2000 years in China: An uncertainty analysis of reconstruction and regional difference," *Geophysical Research Letters*, Volume 37, 2010

Grosjean, Martin, et al., "Ice-borne prehistoric finds in the Swiss Alps reflect Holocene glacier fluctuations," *Journal of Quaternary Science*, Volume 22, Number 3, 2007

Kellerhals, T., et al., "Ammonium concentration in ice cores: A new proxy for regional temperature reconstruction?," *Journal of Geophysical Research*, Volume 115, 2010

Li, H., and Ku, T., "Little Ice Age and Medieval Warm Periods in Eastern China as Read from the Speleothem Records," *American Geophysical Union*, Fall Meeting 2002

Ma, ChunMei, et al., "Climate and environment reconstruction during the Medieval Warm Period in Lop Nur of Xinjiang, China," *Chinese Science Bulletin*, Volume 53, Number 19, 2008

Mantua, Nathan J., et al., "A Pacific interdecadal climate oscillation with impacts on salmon production," *Bulletin of the American Meteorological Society*, Volume 78, 1997

Mantua, Nathan J., "The Pacific-Decadal Oscillation," in *Encyclopedia of Global Environmental Change*, Volume 1, Wiley and Sons, Chichester, 2002

McCabe, Gregory J., "Pacific and Atlantic Ocean influences on multidecadal drought frequency in the United States," *Proceedings of the National Academy of Sciences*, Volume 101, Number 12, 2004

Millar, C.I., et al., "Late Holocene forest dynamics, volcanism, and climate change at Whitewing Mountain and San Joaquin Ridge, Mono County, Sierra Nevada, CA, USA," *Quaternary Research*, Volume 66, 2006

Rutherford, G.K., "Soils of Some Norse Settlements in Southwestern Greenland," *Arctic*, Volume 48, Number 4, 1995

Von Gunten, L., et al., "A quantitative high-resolution summer temperature reconstruction based on sedimentary pigments from Laguna Aculeo, central Chile, back to AD 850," *The Holocene*, Volume 19, 2009

Zhang, De'Er, "Evidence for the existence of the medieval warm period in China," *Climatic Change*, Volume 26, Numbers 2-3, 1994

Zhou, S.Z., et al., "Environmental change during the Holocene in western China on a millennial timescale," *The Holocene*, Volume 1, Number 2, 1991

Chapter 2: A Scandalous Timeout

Lawson, Nigel, *An Appeal to Reason: A Cool Look at Global Warming*, 2008, Overlook Duckworth, Peter Mayer Publishers, Inc.

Lomborg, Bjorn, *Cool It: The Skeptical Environmentalist's Guide to Global Warming*, 2008, Vintage Books

Chapter 3: Giving Shape to Phantoms

Andersen, Geoffrey, "A Rising Tide Swamps All Coasts: What Do We Know About Global Warming," *Slate*, August 15, 2008

Avery, Dennis T., "Look to patterns to grasp glacier growth," *The Journal Star*, December 14, 2008

Booker, Christopher, and North, Richard, *Scared to Death: From BSE to Global Warming – Why Scares are Costing Us the Earth*, 2007, Continuum Books

Canadian Forest Service and National Forestry Database, "Forest Fire Statistics by Province/Territory/Agency, 1970-2009," with data from Ramsey, G.S., and Higgins, D.G., http://nfdp.ccfm.org/data/comp_31e.html

Crichton, Michael, "Aliens Cause Global Warming," Caltech Michelin Lecture, January 17, 2003

Crichton, Michael, *State of Fear*, HarperCollins, 2004

Deseret News, "Climate Expert: Ice Age Coming," September 8, 1972

Feigin, V.L., et al., "A population-based study of the associations of stroke occurrence with weather parameters in Siberia, Russia (1982-92)," *European Journal of Neurology*, Volume 7, 2000

Gething, Peter W., et al., "Climate change and the global malaria recession," *Nature*, Volume 465, 2010

González, Miguel A., and Maidana, Nora I., "Post-Wisconsinian paleoenvironments at Salinas del Bebedero basin, San Luis, Argentina," *Journal of Paleolimnology*, Volume 20, Number 4, 1998

Gray, Louise, "Arctic will be ice-free within a decade," *The Telegraph*, April 7, 2009

Gouveia, N., et al., "Socioeconomic differentials in the temperature-mortality relationship in Sao Paulo, Brazil," *International Journal of Epidemiology*, Volume 32, 2003

Gwynne, Peter, "The Cooling World," *Newsweek*, April 28, 1975

Hansen, James, "Obama's Second Chance on the Predominant Moral Issue of This Century," *The Huffington Post*, April 5, 2010
Hart, Bob, "Current & Recent NCEP/NCAR Reanalysis Snowcover Statistics"

Kievit, Rob, "Faults in climate reports anger Dutch minister," Radio Netherlands Worldwide, February 4, 2010

Kincer, J. B., "Is Our Climate Changing? A Study of Long-time Temperature Trends," *Monthly Weather Review*, Volume 61, Number 9, 1933

Kloner, R.A., et al., "When throughout the year is coronary death most likely to occur? A 12-year population-based analysis of more than 220,000 cases," *Circulation*, Volume 100, 1999

Lander, Mark, "A Look at Global Tropical Cyclone Activity with respect to the Atlantic Change-point Year of 1995"; Proceedings of the 25th AMS Conference

Leake, Jonathan, and Hastings, Chris, "World misled over Himalayan glacier meltdown," *The Times*, January 17, 2010

E. Lioubimtseva, et al., "Impacts of climatic change on carbon storage in the Sahara–Gobi desert belt since the Last Glacial Maximum," *Global and Planetary Change*, Volumes 16-17, 1998

Matthews, Samuel, "What's Happening to Our Climate?" *National Geographic*, November 1976

Office for National Statistics, "Winter Mortality: Excess winter deaths increase in 2008/09," http://www.statistics.gov.uk/cci/nugget.asp?id=574

Onians, Charles, "Snowfalls are now just a thing of the past," *The Independent*, March 20, 2000

Sullivan, Walter, "Expert Says Arctic Ocean Will Soon Be an Open Sea," *The New York Times*, February 20, 1969

United Nations Intergovernmental Panel on Climate Change Fourth Assessment Report: Climate Change 2007, Geneva, Switzerland

Welch, Craig, "Sweeping change reshapes Arctic," *The Seattle Times*, April 9, 2007

Chapter 4: The Vanishing Ice Caps

Bentley, M.J., et al., "Early Holocene retreat of George VI Ice Shelf, Antarctic Peninsula," *Geology*, Volume 33, 2005

Bowles, Claire, "Arctic Meltdown," *New Scientist*, February 27, 2001

Box, J.E., et al., "Greenland ice sheet surface air temperature variability: 1840-2007," *Journal of Climate*, Volume 22, 2009

Chylek, Petr, et al., "Greenland warming of 1920–1930 and 1995–2005," *Geophysical Research Letters*, Volume 33, 2006

Chylek, Petr, et al., "Arctic air temperature change amplification and the Atlantic Multidecadal Oscillation," *Geophysical Research Letters*, Volume 36, 2009

Comte, Michel, "Melting ice reveals ancient hunting tools in Canadian north," Agence France Presse, April 27, 2010

Corr, Hugh F. J., and Vaughan, David G., "A recent volcanic eruption beneath the West Antarctic ice sheet," *Nature Geoscience*, Volume 1, 2008

Danish Meteorological Institute, Daily Mean Temperatures North of 80 Degrees North, http://ocean.dmi.dk/arctic/meant80n.uk.php

R. R. Dickson, et al., "The Arctic Ocean Response to the North Atlantic Oscillation," *Journal of Climate*, Volume 13, 1999

Gerdes, Rüdiger, and Köberle, Cornelia, "Comparison of Arctic sea ice thickness variability in IPCC Climate of the 20th Century experiments and in ocean–sea ice hindcasts," *Journal of Geophysical Research*, Volume 112, 2007

Hanssen-Bauer, Inger, and Førland, Eirik J., "Climate variation in the European Arctic during the last 100 years," *Arctic Climate System Study Final Science Conference: Book of Abstracts*, 2004

Harris, Richard, "Antarctic Ice May Melt, But Not For Millennia," National Public Radio, March 19, 2009

Harris, Richard, "The Mystery of Global Warming's Missing Heat," National Public Radio, March 19, 2008

Imbrie, John, and Imbrie, Katherine Palmer, *Ice Ages: Solving the Mystery*, 1986, Harvard University Press

Kitikmeot Heritage Society, "Hudson's Bay Company,"

http://www.kitikmeotheritage.ca/Angulalk/hudsons/
hudsons.htm

Kitikmeot Heritage Society, "HBC Posts," http://www.
kitikmeotheritage.ca/Angulalk/hudsons/hbcposts/hbcposts.htm

Lansing, Alfred, *Endurance: Shackleton's Incredible Voyage*,
2007, Basic Books

LeMasurier, Wesley E., "Neogene extension and basin
deepening in the West Antarctic rift inferred from
comparisons with the East African rift and other analogs,"
Geology, Volume 36, Number 3, 2008

Lovett, Richard A., "Giant Undersea Volcano Found Off
Iceland," *National Geographic News*, April 22, 2008

Løvø, Gudmund, "Less ice in the Arctic Ocean 6000-7000
years ago," *Norges geologiske undersøkelse*, October 20, 2008

MacDonald, Glen M., et al., "Holocene Treeline History and
Climate Change Across Northern Eurasia," *Quarternary
Research Journal*, Volume 53, 1999

McGhee, Robert, *The Last Imaginary Place: A Human
History of the Arctic World*, 2005, University of Chicago Press

McKay, J.L., et al., "Holocene fluctuations in Arctic sea-ice
cover: dinocyst-based reconstructions for the eastern Chukchi
Sea," *Canadian Journal of Earth Sciences*, Volume 45, 2010

Mysak, L.A., et al., "Sea-ice anomalies observed in the Greenland and Labrador Seas during 1901-1984 and their relation to an interdecadal Arctic climate cycle," *Climate Dynamics*, Volume 5, 1990

Ogilvie, A.E.J., and Jónsdóttir, I., "Sea Ice, Climate, and Icelandic Fisheries in the Eighteenth and Nineteenth Centuries," *Arctic*, Volume 53, Number 4, 2000

Ohashi, Masahiro, and Tanaka, H. L., "Data Analysis of Recent Warming Pattern in the Arctic," *Scientific Online Letters on the Atmosphere*, Volume 6A, 2010

Polyakov, Igor, et al., "Fate of the early-2000s warm water pulse," State of the Arctic Conference, March 2010

President of the Royal Society, *Minutes of Council*, Volume 8, Royal Society, 1817

Mörner, Nils-Axel, "Sea Level Changes and Tsunamis, Environmental Stress and Migration Overseas: The Case of the Maldives and Sri Lanka," *Internationales Asienforum*, Volume 38, 2007

Pisarev, Sergey V., " 'Arctic Warming' During 1920-1940: A Brief Review of Old Russian Publications" in *Report Number 8 on Study of Arctic Change Workshop*, Polar Science Center, Applied Physics Laboratory, University of Washington, 1997

Rigor, Ignatius, and Wallace, John M., "Variations in the Age of Arctic Sea-ice and Summer Sea-ice Extent," *Geophysical Research Letters*, Volume 31, 2004

Sandler, Martin W., *Resolute: The Epic Search for the Northwest Passage and John Franklin, and the Discovery of the Queen's Ghost Ship*, Sterling, 2008

Stenni, B., et al., "The deuterium excess records of EPICA Dome C and Dronning Maud Land ice cores (East Antarctica)," *Quaternary Science Reviews*, Volume 29, 2010

Sundvor, Eirik, et al., "Norwegian-Greenland Sea thermal field," from *Dynamics of the Norwegian Margin*, Nottvedt, et al., editors, The Geological Society of London, 2000

Time magazine staff, "Transport: Northwest Passage II," *Time*, September 13, 1937

Vinther, B.M., et al., "Climatic signals in multiple highly resolved stable isotope records from Greenland," *Quaternary Science Reviews*, Volume 29, 2010

Willerslev, Eske, et al., "Ancient Biomolecules from Deep Ice Cores Reveal a Forested Southern Greenland," *Science*, Volume 317, Number 5834, 2007

Zhang, Jinlun, "Increasing Antarctic Sea Ice under Warming Atmospheric and Oceanic Conditions," *Journal of Climate*, Volume 20, 2007

Zubov, N. N., *Arctic Ice*, translated and edited by U.S. Navy, 1963

Chapter 5: Rise of the Machines

Gerlich, Gerhard, and Tscheuschner, Ralf D., "Falsification Of The Atmospheric CO2 Greenhouse Effects Within The Frame Of Physics," *International Journal of Modern Physics B*, Volume 23, Number 3, 2009

McKitrick, Ross R., and Michaels, Patrick J., "Quantifying the influence of anthropogenic surface processes and inhomogeneities on gridded global climate data," *Journal of Geophysical Research*, Volume 112, 2007

Schwartz, Stephen E., et al., "Why Hasn't Earth Warmed as Much as Expected?," *Journal of Climate*, Volume 23, 2010

Smerdon, Jason E., et al., "Erroneous Model Field Representations in Multiple Pseudoproxy Studies: Corrections and Implications," *Journal of Climate*, in press, 2010

Chapter 6: The Master Narrative

Climate Change 2007, Fourth Assessment Report, Intergovernmental Panel on Climate Change

Fontaine, Russell E., et al., "Malaria Control at Lake Vera, California, in 1952–53, *The American Journal of Tropical*

Medicine and Hygiene, 3(5), 1954

Hecht, Marjorie Mazel, "Where the Global Warming Hoax Was Born," *21ˢᵗ Century Science and Technology*, Fall 2007

Michaels, Patrick, *Meltdown: The Predictable Distortion of Global Warming by Scientists, Politicians, and the Media*, 2004, Cato Institute

Chapter 7: How Hot Is It?

Christy, John R., et al., "Methodology and Results of Calculating Central California Surface Temperature Trends: Evidence of Human-Induced Climate Change," Journal of Climate, Volume 19, 2006

Douglass, David H., and Christy, John R., "Limits on CO_2 climate forcing from recent temperature data of Earth," *Energy and Environment*, Volume 20, Numbers 1 & 2, 2009

Fujibe, Fumiaki, "Urban warming in Japanese cities and its relation to climate change monitoring," *The seventh International Conference on Urban Climate*, 2009

Hansen, James, et al., GISS Global Land-Ocean Temperature Index in 0.01 degrees Celsius, base period: 1951-1980, http://data.giss.nasa.gov/gistemp/tabledata/GLB.Ts+dSST.txt

Hinkel, Kenneth M., and Nelson, Frederick E.,

"Anthropogenic heat island at Barrow, Alaska, during winter: 2001 – 2005," *Journal of Geophysical Research*, VOLUME 112, 2007

Klotzbach, Philip J., et al., "An alternative explanation for differential temperature trends at the surface and in the lower troposphere," *Journal of Geophysical Research*, Volume 114, 2009

Klotzbach, Philip J., et al., "Correction to 'An alternative explanation for differential temperature trends at the surface and in the lower troposphere,' " *Journal of Geophysical Research*, Volume 115, 2010

LaDochy, Steve, et al., "Recent California climate variability: spatial and temporal patterns in temperature trends," *Climate Research*, Volume 33, Number 2, 2007

Lindsey, Rebecca, "Correcting Ocean Cooling," NASA Earth Observatory, November 5, 2008

Lyman, John M., "Recent Cooling of the Upper Ocean," *Geophysical Research Letters*, Volume 33, 2006

McKitrick, Ross R., and Michaels, Patrick J., "Quantifying the influence of anthropogenic surface processes and inhomogeneities on gridded global climate data," *Journal of Geophysical Research – Atmospheres*, December 2007

Montford, Andrew, *The Hockey Stick Illusion: Climategate and the Corruption of Science*, 2010, Stacey International

Petre, Jonathan, "Climategate U-turn as scientist at centre of row admits: There has been no global warming since 1995," *The Daily Mail*, February 14, 2010

Ridley, Matt, "The case against the hockey stick," *Prospect*, March 10, 2010

Streutker, D. R., "A remote sensing study of the urban heat island of Houston, Texas," *International Journal of Remote Sensing*, Volume 23, Number 13, 2002

Willis, Josh K., et al, "Correction to 'Recent Cooling of the Upper Ocean,' " *Geophysical Research Letters, 2007*

Willis, Josh K., et al, "Toward closing the globally averaged sea level budget on Seasonal to interannual time scales," Geophysical Research Abstracts, Volume 10, 2008

Yilmaz, H., et al., "Determination of temperature differences between asphalt concrete, soil and grass surfaces of the city of Erzurum, Turkey," *Atmosfèra*, Volume 21, 2008

Chapter 8: The Quiet Sun

Barron, J.A. and Bukry, D., "Solar forcing of Gulf of California climate during the past 2000 yr suggested by diatoms and silicoflagellates," *Marine Micropaleontology*, Volume 62, 2007

Bonev, Boncho P., et al., "Long-term Solar Variability and the Solar Cycle in the 21st Century," *The Astrophysical Journal*, Volume 605, 2004

Budyko, M. I., "The effect of solar radiation variations on the climate of the Earth," *Tellus*, Volume 21, 1969

Clark, Stuart, "What's wrong with the sun?," *New Scientist*, June 14, 2010

Eddy, John A., et al., *Living With a Star: New Opportunities in Sun-Climate Research*, NASA report, 2003

Eddy, John A., *The Sun, the Earth, and Near-Earth Space: A Guide to the Sun-Earth System*," NASA, 2009

Farrar, Paul D., "Are Cosmic Rays Influencing Oceanic Cloud Coverage Or Is It Only El Niño?" *Climatic Change*, Volume 47, Numbers 1-2, 2000

Fairbridge, Rhodes, and Shirley, James H., "Prolonged Minima and the 179-yr Cycle of the Solar Inertial Motion," *Solar Physics*, Volume 110, 1987

Georgieva, K., et al., "Once again about global warming and solar activity," *Memorie Società Astronomica Italiana*, Volume 76, 2005

Gibson, S. E., et al., "If the Sun is so quiet, why is the Earth ringing? A comparison of two solar minimum intervals," *Journal of Geophysical Research*, Volume 114, 2009

Gran, Rani C., and Layton, Laura, "Space Has Never Been Closer: NASA Instruments Document Contraction of the Boundary between the Earth's Ionosphere and Space," National Aeronautics and Space Administration, December 15, 2008

Levere, Trevor H., *Science and the Canadian Arctic: A Century of Exploration, 1818-1918*, Cambridge University Press, 2004

Lockwood, Mike, et al., "Top-down solar modulation of climate: evidence for centennial-scale change," *Environmental Research Letters*, Volume 5, Number 3, 2010

Long, Marion, "Sun's Shifts May Cause Global Warming," *Discover*, July 2007

Kane, R. P., "A Preliminary Estimate of the Size of the Coming Solar Cycle 24, based on Ohl's Precursor Method," *Solar Physics*, Volume 243, Number 2, 2007

Knudsen, Mads Faurschou, and Riisager, Peter, "Is there a link between Earth's magnetic field and low-latitude precipitation?," *Geology*, Volume 37, Number 1, 2009

Lawrence, E.N., "The Titanic Disaster: a meteorologist's perspective," *Weather*, Volume 55, Part 3

Mackey, Richard, "Rhodes Fairbridge and the idea that the solar system regulates the Earth's climate," *Journal of Coastal*

Research, Special Issue 50, 2007

Muscheler, Raimund, et al., "Long-term climate variations and solar effects," *Solar Variability as an input to the Earth's environment*, International Solar Cycle Studies, Symposium, June 2003

National Aeronautics and Space Agency, "The Sun's Chilly Impact on Earth," December 6, 2001

Ogurtsov, M. G., et al., "Quasisecular cyclicity in the climate of the Earth's Northern Hemisphere and its possible relation to solar activity variations," *Geomagnetism and Aeronomy*, Volume 49, Number 7, 2009

Philips, Tony, "Solar Storm Warning," *NASA Science News*, March 10, 2006
Philips, Tony, "Long Range Solar Forecast: Solar Cycle 25 peaking around 2022 could be one of the weakest in centuries," *NASA Science News*, May 10, 2006

Philips, Tony, "Scientists Predict Big Solar Cycle," *NASA Science News*, December 21, 2006

Philips, Tony, "Solar Wind Loses Power, Hits 50-Year Low," *NASA Science News*, September 23, 2008
Philips, Tony, "Cosmic Rays Hit Space Age High," *NASA Science News*, September 29, 2009

Pustilnik, Lev A., Din, and Gregory Yom, "Influence of Solar

Activity on State of Wheat Market in Medieval England," *Proceedings of International Cosmic Ray Conference 2003*

Rahmstorf, Stevan, et al., "Cosmic Rays, Carbon Dioxide and Climate," *Eos*, January 27, 2004

Saxon, Wolfgang, "Edward P. Ney, 75; Searched the Skies for Cosmic Particles," *The New York Times*, July 12, 1996

Shaviv, Nir J., and Veizer, Jan, "Celestrial driver of Phanerozoic climate?" *Geological Society of America Today*, Volume 13, Number 7, 2003

Shindell, Drew T., et al., "Solar forcing of regional climate change during the Maunder Minimum," *Science*, Volume 294

Steinhilber, F., et al., "Interplanetary magnetic field during the past 9300 years inferred from cosmogenic radionuclides," *Journal of Geophysical Research*, Volume 115, 2010

Svensmark, Hendrik, and Friis-Christensen, Eigil, "Reply to Lockwood and Fröhlich – The persistent role of the Sun in climate forcing," Danish National Space Center, Scientific Report March, 2007

Svensmark, Hendrik, and Enghoff, Martin, "The role of atmospheric ions in aerosol nucleation – a review," *Atmospheric Chemistry and Physics*, Volume 8, 2008

Tinsley, Brian A., and Yu, Fangqun, "Atmospheric ionization

and clouds as links between solar activity and climate," in *Solar Variability and Its Effects on Climate, Geophysical Monograph Series*, Volume 141, edited by J. M. Pap

Usoskin, Ilya G., et al., "Millennium-Scale Sunspot Number Reconstruction: Evidence for an Unusually Active Sun since the 1940s," *Physical Review Letters*, Volume 91, Number 21, 2003

Viereck, Rodney, "The Sun-Climate Connection (Did Sunspots Sink the Titanic?)"; NOAA Space Environment Center, March 26, 2001

Weart, Spencer, "Interview with Jack Eddy," American Institute of Physics, April 21, 1999

Weber, Werner, "Strong signature of the active Sun in 100 years of terrestrial insolation data," *Annalen der Physik*, Volume 522, Number 6, 2010

Willson, Richard C., and Mordvinov, Alexander V., "Secular total solar irradiance trend during solar cycles 21–23," *Geophysical Research Letters*, Volume 30, Number 5, 2006

Xu, Zhentao, "Solar Observations in Ancient China and Solar Variability," *Philosophical Transactions of the Royal Society of London. Series A, Mathematical and Physical Sciences*, Volume 330, Number 1615, 1990

Chapter 9: Unsung Hereos

Carter, Robert, *Climate: The Counter-Consensus*, 2010, Stacey International

Courtillot, V., et al., "Geomagnetic secular variation as a precursor of climatic change," *Nature*, Volume 297, 1982

Courtillot, V., et al., "Are there connections between Earth's magnetic field and climate?," *Earth and Planetary Science Letters*, Volume 253, Issues 3-4, 2007

Evans, David, "No Smoking Hot Spot," *The Australian*, July 18, 2008

Gehrz, Robert D., et al, "Edward Purdy Ney," National Academies Press, 1996

Hantemirov, Rashit M., and Shiyatov, Stepan G., "A continuous multimillennial ring-width chronology in Yamal, northwestern Siberia," *The Holocene*, Volume 12, Number 6, 2002

Kearsley, Geoffrey, "Climate change debate is being distorted by dogma," *Otago Daily Times*, July 17, 2008

Laken, B. A., et al., "Cosmic rays linked to rapid mid-latitude cloud changes," *Atmospheric Chemistry and Physics*, Volume 10, 2010

Lindzen, Richard, "Climate Science: Is it currently designed to answer questions?" paper given at a 2008 conference entitled "Creativity and Creative Inspiration in Mathematics, Science, and Engineering: Developing a Vision for the Future"

Martin G. Mlynczak, et al, "First light from the Far-Infrared Spectroscopy of the Troposphere (FIRST) instrument," *Geophysical Research Letters*, Volume 33, 2006

Montford, A. W., *The Hockey Stick Illusion*, Stacey International, 2010

Palle, E., et al., "The possible connection between ionization in the atmosphere by cosmic rays and low level clouds," *Journal of Atmospheric and Solar-Terrestrial Physics*, Volume 66, 2004

Singer, S. Fred, and Avery, Dennis T., *Unstoppable Global Warming: Every 1,500 Years*, 2007, Rowman & Littlefield Publishers, Inc.

Solomon, Lawrence, *The Deniers: The World-Renowned Scientists Who Stood Up Against Global Warming Hysteria, Political Persecution, and Fraud*, 2008, Richard Vigilante Books

Svensmark, Henrik, and Calder, Nigel, *The Chilling Stars: A New Theory of Climate Change*, 2007, Icon Books

Thompson, Andrea, "Earth's Clouds Alive With Bacteria," LiveScience, February 28, 2008

Tinsley, Brian A., "Influence of Solar Wind on the Global Electric Circuit, and Inferred Effects on Cloud Microphysics, Temperature, and Dynamics in the Troposphere," *Space Science Reviews*, Volume 94, Issue 1/2, 2000

Chapter 10: Euphoric Recall

Ahlbeck, Jarl R., "No significant global warming since 1995," *Facts & Arts*, October 25, 2008

Alley, R. B., "The Younger Dryas cold interval as viewed from central Greenland," *Quaternary Science Reviews*, Volume 19, 2000

Angell, J. K., and Korshover, "Estimate of the Global Change in Temperature,Surface to 100 mb, Between 1958 and 1975," *Monthly Weather Review*, Volume 105, 1977

Bryson, Bill, *A Short History of Nearly Everything*, Broadway, 2004

Christy, John R., and Hnilo, Justin J., "Changes in Snowfall in the Southern Sierra Nevada of California since 1916," *Energy and Environment*, Volume 21, No. 3, 2010

Chu, Guoqiang, et al., "The 'Mediaeval Warm Period' drought recorded in Lake Huguangyan, tropical South China," *The Holocene*, Volume 12, Number 5, 2002

Del Genio, Anthony, "Separating the Man-Made from the

Natural," Goddard Institute for Space Studies, December, 2008

Fagan, Brian, *The Little Ice Age: How Climate Made History 1300-1850*, 2000, Basic Books

Griggs, Kim, "Big ice shelf's disappearing act," BBC News, December 4, 2006

Hobgood, Jay S., and Cerveny, Randall S., "Ice-age hurricanes and tropical storms," *Nature*, Volume 333, 1988

Joerin, Ulrich A., et al., "Multicentury glacier fluctuations in the Swiss Alps during the Holocene," *The Holocene*, Volume 16, Number 5, 2006

Kaufman, D.S., et al., "Holocene thermal maximum in the western Arctic (0–180 W)," *Quaternary Science Reviews*, Volume 23, 2004

Keys, David, *Catastrophe: An Investigation into the Origins of the Modern World*, 1999, Ballantine

Landsea, Christopher W., et al, "Impact of Duration Thresholds on Atlantic Tropical Cyclone Counts," *Journal of Climate*, Volume 23, Issue 10, 2010
Lippsett, Lonny, "Ocean Conveyor's 'Pump' Switches Back On," *Oceanus*, Woods Hole Oceanographic Institution, January 9, 2009

Mangini, A., et al., "Reconstruction of temperature in the

Central Alps during the past 2000 years from an Oxygen-18 stalagmite record," *Earth and Planetary Science Letters*, Volume 235, July 2005

Mehta, Vikram M., et al., "Oceanic influence on the North Atlantic Oscillation and associated Northern hemisphere climate variations," *Geophysical Research Letters*, Volume 27, January 2000

Mowatt, Farley, *Lost in the Barrens*, 1982, Bantam Books

Mowatt, Farley, *The Snow Walker*, 2004, Stackpole Books

Revkin, Andrew, "Gore Pulls Slide of Disaster Trends," *The New York Times*, February 23, 2009

Reyes, Alberto V., et al, "Expansion of alpine glaciers in Pacific North America in the first millennium A.D.," *Geology*, Volume 34, Number 1, 2006

Rodwell, M.J., et al., "Oceanic forcing of the wintertime North Atlantic Oscillation and European Climate," *Nature*, Volume 398, March 25, 1999

Rodwell, M.J., and Folland, C.K., "*Atlantic air-sea interaction and seasonal predictability*," *Quarterly Journal of the Royal Meteorological Society*, Volume 128, Issue 583

Willis, Josh, et al., "Toward closing the globally averaged sea level budget on Seasonal to interannual time scales,"

Geophysical Research Abstracts, Volume 10, 2008

Zhang, David D., et al., "Global climate change, war, and population decline in recent human history," *Proceedings of the National Academy of Sciences of the United States of America*, 2007

Chapter 11: A Broken Moral Compass

Barnes, Deborah Corey, "The Money and Connections Behind Al Gore's Carbon Crusade," *Human Events*, October 3, 2007

Booker, Christopher, "The 'Consensus' on Climate Change is a Catastrophe Itself," *The Telegraph*, August 31, 2008
Bryce, Robert, "Windmills Are Killing Our Birds," *The Wall Street Journal*, September 7, 2009

Cohen, Martin, "Beyond debate?," *Times Higher Education*, December 10, 2009

Congress of Racial Equality, "House Climate Bill Called 'Immoral' by Major Civil Rights Leader," June 25, 2009, http://enews.core-online.us/mail/util.cfm?gpiv=2100041 936.15518.285&gen=1

Foreman, Jonathan, "Taking the private jet to Copenhagen," *The Times*, November 29, 2009
Galbraith, Kate, "Ice-Tossing Turbines: Myth or Hazard?,"

The New York Times, December 9, 2008

Goodrich, Craig, "Green as in Money," *The Libertarian Enterprise*, Number 573, June 6, 2010

Harrabin, Roger, "UK expands wind power potential,' " BBC News, June 24, 2009

Harrabin, Roger, "Wind 'can revolutionise UK power,' " BBC News, July 1, 2009,

Hedge, Aaron, "McKibben calls on Aspen Ideas audience to take action now," *The Aspen Times*, July 12, 2010

Horner, Christopher, "Something Rotten in the NYT," *National Review Online*, March 27, 2008

Klaus, Vaclav, *Blue Planet in Green Shackles*, 2007, Competitive Enterprise Institute

Leake, Jonathan, "The great climate change science scandal," *The Times*, November 29, 2009

Lyttle, Steve, "Gore gets a cold shoulder," *The Sydney Morning Herald*, October 14, 2007

Prather, Michael J., and Hsu, Juno, "NF3, the greenhouse gas missing from Kyoto," *Geophysical Research Letters*, Volume 35, 2008

Sierra Club, Kansas Chapter, "Fact Sheet on Mercury Pollution from Coal-Fired Power Plants in Kansas," http://kansas.sierraclub.org/Wind/Coal-MercuryFactSheet.htm

Soon, Willie, and Driessen, Paul, "Eco-colonialism Degrades Africa," Science & Public Policy Institute, Commentary and Lecture Series, February 14, 2009

U.S. Army Corps of Engineers, "Marinas on Center Hill Lake," http://www.lrn.usace.army.mil/op/cen/rec/marinas.htm

Weston, Luke, "Nitrogen trifluoride as an anthropogenic-greenhouse-forcing gas," Physical Insights, July 3, 2008

Chapter 12: Don't Sell Your Coat

Booker, Christopher, "The world has never seen such freezing heat," *The Telegraph*, November 16, 2008

Borland, John, "Lunar Eclipse Prompts Climate Change Debate," *Wired Science*, March 3, 2008

Boyes, Roger, "Iceland prepares for second, more devastating volcanic eruption," *The Times*, March 21, 2010

Brahic, Catherine, "Thousand of new volcanoes revealed beneath the waves," *New Scientist*, July 9, 2007

CTV News, "Cruise ship freed from St. Lawrence River ice," January 27, 2009

Dunk, Marcus, "Will Krakatoa rock the world again? Last time, it killed thousands and changed the weather for five years, now it could be even deadlier...," *The Daily Mail*, July 31, 2009

Easterbrook, Donald J., "Solar Influence on Recurring Global, Decadal, Climate Cycles Recorded by Glacial Fluctuations, Ice Cores, Sea Surface Temperatures, and Historic Measurements Over the Past Millennium," *American Geophysical Union Fall Meeting 2008*

Fairbridge, Rhodes W., et al., "Solar-Planetary-Climate Stress, Earthquakes, and Volcanism," NASA-funded research, 1979

Grad, Shelby, "L.A.'s cool summer continues: Lancaster posts a record low temperature," *The Los Angeles Times*, August 18, 2009

Hillier, J. K., and Watts, A. B., "Global distribution of seamounts from ship-track bathymetry data," *Geophysical Research Letters*, Volume 34, 2007

Kessell, Ramona, et al., "ULF energy transfer in the solar wind - magnetosphere - ionosphere - solid Earth system," Geophysical Research Abstracts, Volume 8, 2006

Lu, Qing-Bin, "What is the Major Culprit for Global Warming: CFCs or CO_2?," *Journal of Cosmology*, Volume 8, 2010

Lyttle, Steve, "Gore Gets a Cold Shoulder," *The Sydney Morning Herald*, October 14, 2007

Medred, Craig, "Bad Weather Was Good for Alaska Glaciers," *The Anchorage Daily News*, October 13, 2008

Minard, Anne, "Sun Oddly Quiet -- Hints at Next "Little Ice Age"?," *National Geographic News*, May 4, 2009

PDO Index, Joint Institute for the Study of the Atmosphere and Ocean, University of Washington, http://jisao.washington.edu/pdo/PDO.latest

Rampino, Michael R., "Can Rapid Climatic Change Cause Volcanic Eruptions?," *Science*, Volume 206, Number 4420, 1979

Schneider, David, et al., "Climate response to large, high-latitude and low-latitude volcanic eruptions in the Community Climate System Model," *Journal of Geophysical Research*, Volume 114, 2009

Shearing, Caroline, "Ski resorts open early after heavy snow," *The Telegraph*, November 11, 2008

Stewart, Will, "Russia's top weatherman's blow to climate

change lobby as he says winter in Siberia may be COLDEST on record," *The Daily Mail*, March 24, 2010

Strestik, Jaroslav, "Possible Correlation Between Solar and Volcanic Activity in a Long-term Scale," *Solar variability as an input to the Earth's environment. International Solar Cycle Studies (ISCS) Symposium*, 2003

. . . .

Photo Credits

By Page Number. Clockwise from top where more than one image per page.

* * * * * * * * * * * * *

1 Library of Congress

2 Ponting, Herbert George; Library of Congress

6 Mike Dunn; NOAA Climate Program Office, NABOS 2006 Expedition

10 Photo courtesy of Habibullo Abdussamatov

12 Photo courtesy of Habibullo Abdussamatov

14 Alan Wilson; http://www.naturespicsonline.com/copyright

22 Based on data from J. R. Petit, et al.

23 Library of Congress

24 United States Federal Government/Wikipedia Commons

25 NASA image by Jesse Allen, AMSR-E data processed and provided by Chelle Gentemann and Frank Wentz, Remote Sensing Systems

26 NASA/Jet Propulsion Laboratory

32 NASA, Visible Earth Project

40 Army Corps of Engineers

43 Library of Congress

54 NASA's Earth Observatory/Robert Simmon and GeoEye

63 Library of Congress

69 Randall Osterhuber, University of California-Berkeley, Central Sierra Snow Laboratory, http://research.chance.berkeley.edu

76 Michael Van Woert, National Oceanic and Atmospheric Association

84 National Snow and Ice Data Center

88 National Snow and Ice Data Center

98 NASA

110 Historic National Weather Service Collection

114 Wikipedia Commons

118 Library of Congress

130 *Climate Change 2001: The Scientific Basis.* Contribution of Working Group I to the Third Assessment Report of the Intergovernmental Panel on Climate Change, Summary for Policymakers, Figure 1. Cambridge University Press.

135 George Taylor, surfacestations.org

139 NASA

142 Greg Rico/Penn State; House of Representatives

144 *Climate Change 2001: The Scientific Basis.* Contribution of Working Group I to the Third Assessment Report of the Intergovernmental Panel on Climate Change, Summary for Policymakers, Figure 1. Cambridge University Press.

145 From *Taken By Storm*, Christopher Essex and Ross McKitrick

146 Climatic Research Unit, University of East Anglia

147 NASA/Goddard Institute for Space Studies

149 NASA/Goddard Institute for Space Studies

150 Data from National Oceanic and Atmospheric Administration

152 SiriusB

155 Solar and Heliospheric Observatory/NASA

159 Recreated graph based on NASA original

162 Barbara Eddy

168 William Livingston and Matthew Penn; Rob Rutten

169 William Livingston and Matthew Penn

172 Lars Oxfeldt Mortensen, from his film The Cloud Mystery, Photo copyright Mortensenfilm.dk

175 Simon Swordy/NASA

177 Sodankyla Geophysical Observatory, University of Oulu

192 Public Domain

194 Ryan Maue

203 Wikipedia Commons

210 Library of Congress

213 Department of Agriculture; Sakurai Midori; Christian Fischer

226 NASA's Earth Observatory

233 Painting by Reginald Easton; Public Domain

Made in the USA
Lexington, KY
17 March 2012